THE LITTLE BOOK OF COSMOLOGY

the

LITTLE BOOK

of

COSMOLOGY

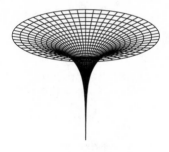

L YMAN P AGE

PRINCETON UNIVERSITY PRESS   PRINCETON AND OXFORD

Requests for permission to reproduce material from this work
should be sent to permissions@press.princeton.edu

Published by Princeton University Press
41 William Street, Princeton, New Jersey 08540
6 Oxford Street, Woodstock, Oxfordshire OX20 1TR

press.princeton.edu

ISBN 978-0-691-19578-0
ISBN (e-book) 978-691-20169-6

British Library Cataloging-in-Publication Data is available

Editorial: Ingrid Gnerlich and Arthur Werneck
Production Editorial: Brigitte Pelner
Text and Cover Design: Jessica Massabrook
Production: Jacqueline Poirier
Publicity: Sara Henning-Stout (US) and Katie Lewis (UK)
Copyeditor: Karen Verde

This book has been composed in Bembo Std

Printed on acid-free paper ∞

Printed in the United States of America

1 3 5 7 9 10 8 6 4 2

To Lisa and the boys

CONTENTS

	Preface	*ix*
	Acknowledgments	*xv*
CHAPTER ONE	The Basics	1
CHAPTER TWO	The Composition and Evolution of the Cosmos	28
CHAPTER THREE	Mapping the Cosmic Microwave Background	58
CHAPTER FOUR	The Standard Model of Cosmology	81
CHAPTER FIVE	Frontiers of Cosmology	98
	Appendixes	*111*
	Index	*119*

PREFACE

These pages provide a brief introduction to modern cosmology, the study of the universe at the most extreme scales of space, energy, and time. My hope is that they convey some essential aspects of what we know about the universe—including its composition, its geometry, its evolution, and the laws of physics that describe it—and how we know it. The subject of the universe is ripe for wild theories and speculation, but these hide perhaps its most amazing aspect: we can understand the universe at its grandest scales to percent-level accuracy through measurement.

As we shall see, the universe at the largest scales and earliest times is remarkably simple and can be characterized with just a few parameters. It is much easier to understand than, say, the fascinatingly complex Earth, with its atmosphere, oceans, moving continents, and magnetic field, to name just a few attributes. In this book, I will try to explain not only current observations and measurements, but also how they can be woven together via physical explanations into a unified picture of the cosmos. The picture I'll describe is not the only possible one, but it explains the data with a minimal set of assumptions. Continuing observations will reveal whether it is correct.

Our knowledge of the universe is encapsulated in what we call the Standard Model of Cosmology, and it agrees remarkably well with observations. It is predictive, testable, and could easily be falsified or augmented if that were called for. Among other things, the model says that the universe is comprised of about 5% atomic material, the stuff of which we are made; about 25% "dark matter"; and 70% "dark energy." Based on Einstein's theory of gravity, the Standard Model specifies how the various components of the universe evolve from the very earliest times to the present. Put another way, we take from general relativity a way of thinking about space and use it as a foundation for describing how the cosmic components—the radiation, atoms, dark matter, and dark energy—fit together to make the universe we observe. All this said, although we have an excellent model of the universe, we do not yet have a fundamental understanding of its dominant constituents. There are exciting open questions in cosmology that continue to be investigated by scientists throughout the world, and we will cover some of them at the end of this book.

Following my own path to learning about cosmology, the focus of the book will be on understanding the universe through measurements of the Cosmic Microwave Background (CMB)—the faint thermal afterglow of the birth of our universe. The evidence in support of this interpretation is overwhelming. Though the CMB resembles radiant heat from the Sun or from an electric stove burner, it has a much, much colder temperature. Hinting at its ancient origin, it is a mere $2.725°C$ above absolute zero, or 2.725 K.[1] But, there is

---

1   The CMB is often called the "3K background" because 2.725 K is almost 3 K. The number of °C above absolute zero corresponds to the kelvin temperature scale. That is, 1°C above absolute zero is 1 K; there is no "°" sign for kelvin. A change of, say, 0.01°C is the same as a change of 0.01 K. In this system, which we will use

much more to the CMB than just its temperature. Indeed, most of what we learn from it comes from tiny variations in its temperature from position to position across the sky. For example, the CMB is ever so slightly different in temperature in (to pick two arbitrary directions) the north and south celestial poles. Because the CMB can be measured in such exquisite detail, our understanding of it is the foundation for our cosmological model. However, before we delve into the various characteristics of the CMB and what they tell us, we will first need to develop some basic ideas of how to think about the universe as a whole.

In chapter 1, we will lay our foundations and cover some basics about the cosmos, guided by two observations: that the speed of light is finite and that the universe is expanding. The meshing of these two facts creates a framework that we will use in subsequent chapters. In chapter 2 we will review the composition of the universe, not in minute detail but rather, focusing on the components that dominate in different epochs of our cosmic history. It is the composition of the universe that tells it how to evolve. We will also discuss how the components of the universe work together to form stars, galaxies, and clusters of galaxies. In cosmology these are simply called "structure." The whole process of structure formation is rooted in the Big Bang and ultimately gave rise to Earth and, eventually, to us. In chapter 3 we will explain the tiny temperature variations in the CMB that are shown in plate 1. Through understanding this image we can understand a tremendous amount about the universe. Then

---

from here on, absolute zero is $-273.14°$C, water freezes at $0°$C or 273.14 K, and boils at $100°$C or 373.14 K. The Sun is about $5500°$C or 5773 K, which we will approximate as 6000 K from here on.

in chapter 4, we will bring the pieces together and present the Standard Model of Cosmology. Though the standard model is wonderfully predictive, much remains unknown. Finally, in chapter 5, I'll describe some of the frontiers of theoretical and experimental research in cosmology.

Cosmology is a vibrant and exciting field. The search for ever deeper knowledge on both theoretical and experimental fronts is ongoing. For observers of the cosmos, such as myself, the CMB continues to offer insights—and continued measurements may yet lead us to look at elements of the standard model in a new light, and may also guide us to new discoveries.

Before we begin, let me add a brief note about the level of this book. One of the challenges in presenting recent developments in science is to pitch ideas at the right level for the reader. While I define various terms and concepts with scientific specificity, throughout the book I do make some assumptions about the reader's background knowledge and inherent level of interest. This said, I have added a few appendixes to provide a little more detail on certain topics, if needed. For example, I assume that readers will know that light is a wave of a certain wavelength that carries energy; however, appendix A.1, entitled, "The Electromagnetic Spectrum," provides a short guide to various sources of radiation[2] and their wavelengths, in case the reader wants further information on the subject. Also, I expect that most readers will know that the speed of light is finite and a fundamental constant of Nature. However, what is less widely appreciated is that, no matter where you are in the universe or how fast you are moving, you will measure that the speed of light in a vacuum is

---

2   I'll use the terms "light" and "radiation" synonymously.

186,000 miles per second. This is one of the foundations of Einstein's special theory of relativity. To keep this book concise, I will not delve too deeply into relativity (there are many books that do this well already) or other such topics; but, as we go along, I will explain physical concepts related to our understanding of the cosmos in a little more detail than you may have encountered in the past. By necessity, I will be somewhat quantitative, but rest assured that the math needed will be at the level of *distance = speed × time*; and most of the time, we will use approximate numbers as they are easier to grasp.

A tricky element in cosmology is that the distances and timescales are so large they can be difficult to imagine. To make them easier to comprehend, we will count things in "billions." To put this number in some context, there are somewhat more than seven billion people on Earth; the tip of your little finger contains about one billion cells; and one billion M&Ms would slightly overfill a cubic box about six meters on a side. Since this is a popular-level book (and hoping my colleagues forgive me), there are no scientific references, and the attribution of specific ideas and findings is minimal.

There is a lot to cover in this short book—a whole universe—so let's get going!

ACKNOWLEDGMENTS

I HAVE HAD THE GOOD FORTUNE TO LEARN COSMOLOGY from many of the leaders in theory. David Spergel has been a close collaborator for over two decades. Jim Peebles and Paul Steinhardt have answered many questions; both made critical suggestions for this book. Dick Bond has tutored me since I was a postdoc. Slava Mukhanov taught me about the early universe.

Of course, all the errors are mine: one can always be a better student. Both Steve Boughn and Shyam Khanna read earlier versions in detail and made multiple suggestions that I have included. Jeff Aumuller advised on how to make it understandable as have Kevin Crowley, Oriel Farajun, Bryant Hall, Neha Anil Kumar, Loki Lin, Christian Robles, Allan Shen, Mona Ye, and Kasey Wagoner. Ingrid Gnerlich, my editor, gave the book its current form, made suggestions too numerous to mention, and was a delight to work with.

A special acknowledgement is due to my colleague, Steve Gubser, who initiated a series of "little books" in physics—one on string theory and another on black holes, with Frans Pretorius—and thereby set out the footsteps in which this book follows. Steve tragically died while rock climbing in 2019. This book is one of the many ways in which he will be remembered.

THE LITTLE BOOK OF COSMOLOGY

CHAPTER ONE

THE BASICS

1.1 *The Size of the Universe*

HOW BIG IS THE UNIVERSE? IT IS REALLY, REALLY BIG! MORE seriously, this is a deep question. Addressing it will take us to the heart of cosmology. However, before we get to what the question even means, let us first consider some typical distances. In cosmology, distances are truly vast. To set the scale we will start locally and then work our way out. The Moon is about 250,000 miles away and is considered nearby. Its distance is close to the typical mileage on a car before it breaks down. With a really good car you could imagine driving to the Moon and possibly even making it back. However, if we go beyond the Moon, it becomes cumbersome to keep measuring distances in miles. Because the universe is so vast, we typically measure distances another way—with light. We can ask how long it takes light to travel from an object to us. Since the speed of light is a constant of Nature, it is a

2

convenient standard. In one second light travels 186,000 miles. Put another way, one light-second is the distance light travels in one second (186,000 miles). Similarly, in 1.3 seconds, light travels 250,000 miles. Now, instead of specifying miles, we can say the Moon is 1.3 light-seconds away. Note that we are using a time-like term (light-seconds) to talk about distance.

The Sun is on average about 93 million miles from us, or about eight light-minutes away.[1] Because the fastest speed at which information can travel is the speed of light, when something happens on the surface of the Sun we must wait about eight minutes for the light from the event to reach our eyes. We will revisit this concept, applied to the cosmic scale. For now, though, we will focus on distances and not on the time it takes to travel that distance.

The next time you are away from city lights on a moon-less night and look up at the night sky, you will see a swath that is brighter than everything else. This glow comes from billions of stars that are part of the Milky Way, our galaxy, of which our Sun is a fairly typical star. A typical galaxy contains roughly one hundred billion stars. One way to connect with this number is that our brains have about one hundred billion neurons; so, there is a neuron in your brain for every star in our galaxy.

The stars in the Milky Way are collected in a sort of disc shape that is about 100,000 light-years in diameter and has a bulge in the middle. Figure 1.1 shows a sketch of how it might appear if we could view the Milky Way from a distance. The galactic plane is an imaginary surface that cuts the disc in half as though you were slicing a hole-less bagel. The solar system

---

1   The distance of 93 million miles divided by 186,000 miles per second is 500 seconds, or a little more than eight minutes.

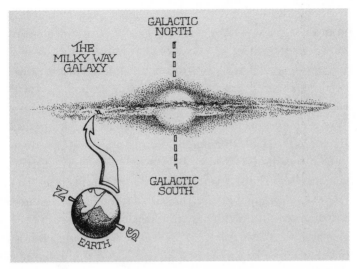

GALACTIC
NORTH

THE
MILKY WAY
GALAXY

GALACTIC
SOUTH

EARTH

FIGURE 1.1. The Milky Way as seen by an imaginary viewer at a distance. The overall shape resembles a disc with a bulge in the middle. The galactic center is at the middle of the bulge. The orientation of the Earth with respect to the galaxy is approximate. *Credit:* Stewart Brand and Jim Peebles in The CoEvolution Quarterly.

is about halfway out from the center of the disc. When we look toward the center of the disc, we see many more stars than when we look well off to the side. It is a bit like living on the outskirts of a city. You are a part of the city, but you can still see all of the tall buildings off in one direction.

Plate 2 is a picture of the Milky Way, taken with a CCD camera using visible light.[2] If our eyes were more sensitive

2 Our eyes detect the spectrum of colors that make up visible light, each color corresponding to a different wavelength. A typical visible wavelength is about one hundredth the thickness of a human hair. More formally, a this typical wavelength is 0.5 microns, where one micron is a thousandth of a millimeter . There are many other possible wavelengths of light that we cannot see. Taken together, these are called the "electromagnetic spectrum," as shown in appendix A.1.

THE BASICS

and larger, we would see the galaxy like this. The dark swaths in this image come from dust in our galaxy that obscures the starlight, somewhat like smoke obscuring flames from a fire. In cosmology, "dust" refers to microscopic particles comprised of a variety of materials including carbon, oxygen, and silicon. Plate 3 shows a different view of the Milky Way, this one made by the Diffuse InfraRed Background Explorer (DIRBE), an infrared telescope and one of the three instruments on the COsmic Background Explorer (COBE), satellite.[3] Unlike the image in plate 2, this was made at "far-infrared" wavelengths, in particular at 100 microns. Infrared radiation tells us how things emit heat. In this image we see primarily the thermal glow of the Milky Way, in other words, the emission of heat. The heat comes from the dust that fills our galaxy, the same dust that obscures the starlight.

A typical galaxy like the Milky Way has an average temperature of about 30 K, so it is not very hot but it still emits thermal energy. We can draw a loose analogy with an incandescent lightbulb. The bulb is most obvious to us because of the visible light it emits, analogous to the light in plate 2. However, the lightbulb produces much more energy as heat that we can feel but cannot see.[4] When you touch an incandescent bulb it is hot. You may have seen pictures of houses taken in infrared light. These pictures tell you where the heat is leaking out

3 The other two instruments discovered the anisotropy in the CMB (DMR, the Differential Microwave Radiometer, leader George Smoot) and made the definitive measurement of CMB temperature (FIRAS, the Far InfraRed Absolute Spectrophotometer, leader John Mather). Mike Hauser led DIRBE. The instrument is best known for detecting the combined thermal emission of all the galaxies in the universe.

4 Modern LED or CFBs have a higher ratio of visible light to heat, which is why they are more efficient for lighting.

CHAPTER ONE

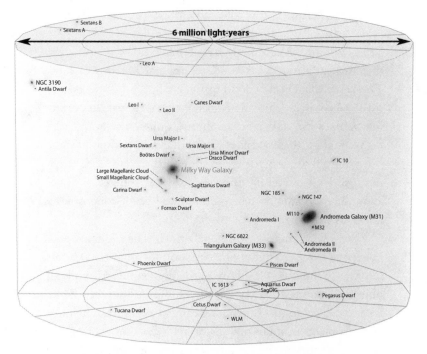

FIGURE 1.2. The Local Group of galaxies. Andromeda is about 2.5 million light-years away but can be seen with the naked eye in dark conditions away from city lights. In length it appears a few times as large as the full moon. The Magellanic Clouds are readily visible by eye in the southern hemisphere. The larger one, close to the Milky Way in this image and shown in plate 3 emitting thermal radiation, is about twenty full moons across. The top and bottom wire grid "wheels" are six million light-years in diameter. *Credit:* Andrew Z. Colvin, https://en.wikipedia.org/wiki /Local_Group. Need formal permission.

(often at the windows). When you feel the heat from a hot body, it is mostly infrared radiation that you sense.

Let's take another step out into the cosmos. Our galaxy is a member of the "Local Group" of roughly 50 galaxies, as shown in figure 1.2. The Local Group is some six million light-years across. In this collection, the Milky Way is second in size to the Andromeda galaxy but the range of sizes is quite large. Whereas Andromeda has about a 1,000 billion stars, the smaller "dwarf" galaxies have tens of millions of stars. The Large Magellanic Cloud (plate 3 & figure 1.2) is a nearby small galaxy that orbits the Milky Way.[5] With galaxies orbiting galaxies the distances are already quite large but, as the name implies, these galaxies are still "local." Although there is no sharp boundary for when something is said to be "cosmological," we typically think in terms of spheres or cubes about 25 million light-years across. The Local Group is just a fraction of this size.

Plate 4 is an amazing image, taken with the Hubble Space Telescope by observing in one direction for almost 300 hours in order to build up sensitivity to the light emitted from faint objects. The image, known as the Hubble Ultra Deep Field, is somewhat akin to a super-long camera exposure. The most distant objects in it are billions of light-years away. The area covered by the image is about a sixtieth the area of the full moon. We can be a bit more quantitative. The angular width of the full moon is about one-half a degree across, or roughly half the size in angle of your little finger when held up at arms length.[6] You can compute that it takes 200,000 full moons to cover the full sky. Here is the mind-blowing thing about the image: only a handful of the objects in it are stars—the large

5   Although named after Ferdinand Magellan's report of them in 1519, they were first recorded more than 500 years earlier by Abd al-Rahman al-Sufi Shirazi, a Persian astronomer.
6   If you lined up full moons side by side, it would take 720 to make a circle that went through both the north and south celestial poles (or any great circle) because there are 360 degrees in a circle. The conventional notation is that the Moon is 0.5° in diameter, or, in our example, 360°/720.

CHAPTER ONE

majority of objects are *galaxies*. And each of those galaxies typically includes about 100 billion stars.

To determine the number of galaxies in the image, you simply need to count them. With a full-resolution picture you could do this by hand, but it is easier to use computers. The Hubble Ultra Deep Field team finds about 10,000 galaxies in the image, which means that across the full sky there are about 100 billion galaxies.[7] To emphasize, we *observe* that there are a *finite* number of typically sized galaxies. We say that in the *observable universe*, the subset of the whole universe that is observable by us in principle, there are roughly 100 billion galaxies, each typically with about 100 billion stars. It is a coincidence that the numbers are so close.

We have just introduced a profound concept, that of the "observable universe," and a profound observation, that in the Hubble Ultra Deep Field we have observed essentially all the Milky Way type galaxies that can be seen in that direction. In other words, with the Hubble Ultra Deep Field we have gone as far as we can in counting objects. To understand these ideas, we will have to consider a universe that evolves with time, as we do below, but first we want to continue to think of the universe as an endless and static expanse that we can explore at will.

If we could freeze time and tour the universe, what would we see? Let's put aside the finite speed of light and imagine that someone, say Alice, could go anywhere in the universe instantaneously and communicate with someone else

---

7   This is [10,000 galaxies per Hubble Ultra Deep Field]×[60 Deep Fields per full Moon]×[200,000 full Moons in the full sky]=120,000,000,000 which we round down to 100 billion. If we looked at much lower mass galaxies than are readily seen with the Hubble, we might get a factor of 10 more, but each would have far fewer stars.

instantaneously. We can think of galaxies as cosmic signposts. We can, in principle, give them names and know where they are in the universe. As you can see in the image of the Local Group in figure 1.2, this accounting has already been done locally. But we want to go to much greater distances. Let's say Alice is in a distant galaxy that is ten billion light-years away. We ask her to describe the local cosmic environment in broad terms, such as the number and general appearance of the other galaxies near her. We then compare our description from our home in the Milky Way to Alice's. We find the descriptions are similar. Although there would be a large variety of galaxies, no matter where we went, no matter how far away, no matter what direction, on average the galactic environment would look very much like it does right around us, and the same laws of physics would describe Nature.

This is an important conceptual point and is worth repeating because we will build on it. At this instant in time, every place in the universe looks, in broad brush strokes, similar. We could call up someone near any distant galaxy and ask them to describe the galaxies within a 25 million light-year diameter sphere centered on them. We would find that their general description also described our galactic neighborhood.

The idea that the universe is on average the same everywhere you go at a specific time is called Einstein's "cosmological principle." When a quantity is similar everywhere in space, it is said to be *homogeneous*. The cosmological principle thus says that the universe is homogeneous when averaged over a large enough volume. The cosmological principle also says that on average, the universe looks the same in each direction. This property is called *isotropy*. It means that on average the picture from the Hubble Ultra Deep Field would look the same regardless of the direction we pointed the satellite as long as we looked away from nearby objects like the galactic plane. Our universe is homogeneous and isotropic no matter where we are in it.

The concepts of homogeneity and isotropy are related but distinct. For example, if your universe were a grapefruit and you lived at its center, you'd say your cosmology was isotropic (ignoring the membranes around the pulp), but because the pulp is in the middle and the rind is on the outside, you would say it is not homogeneous. It took a conceptual advance to postulate the cosmological principle. In our day-to-day lives the sky is far from isotropic: we see the Sun rise and set, and the solar system is far from homogeneous as the planets lie roughly in a plane. To think about the universe, we need to step away and imagine a much more simple distribution of matter on a much, much larger scale.

We have completed a whirlwind tour of the universe. We stepped out to greater and greater distances until, with the Hubble Ultra Deep Field, we ran out of objects to observe. To understand how this can happen, we will need to consider the evolution of the universe in time, which we do in the following sections. That aside, limiting ourselves to a purely spatial description, we got out far enough to envision a homogeneous universe frozen in time and full of galaxies, on average, like the ones around us. At this instant, we can think of the universe as an endless three-dimensional grid of Tinkertoys with the hubs representing collections of galaxies that look in general like the ones around us. Of course, the galaxies are distributed throughout space and not on a grid pattern, but the Tinkertoys help us imagine a coordinate system for describing the cosmos.

1.2 The Expanding Universe

In the last section we imagined the universe as static, but it is not: the universe is expanding. This is not a theory, or a model—it is an observational fact. Once we get well past the Local Group (figure 1.2) and out to cosmological distances, we observe that *the farther away a galaxy is, the faster it is moving away from us.* This is called the Hubble-Lemaître Law, after Georges Lemaître who, based on observations available at the time, published it in an obscure journal in 1927, and Edwin Hubble, who published it independently in 1929. In day-to-day terms, the Hubble-Lemaître law states that for every million light-years away you observe an object, its recessional speed increases by about 15 miles per second. This value is called "Hubble's constant."[8]

Hubble's observation immediately brings to mind the question: Are we at the center of the universe? The answer is no. Just because we see all galaxies rushing away from us, it does not mean that we are at the center of the universe. We are special but not *that* special. All observers on all galaxies any-where in the observable universe see the same thing. This is because the expansion has a particular form, namely that the recessional speed is proportional to the distance. That is, if a galaxy is twice as far away, it is moving away from us twice as fast. Let's be more concrete and imagine a sample string of galaxies each representing their local region of 25 million

---

8 Hubble's original value, which was based on his observations of distances and velocities measured by Vesto Slipher, was about seven times the currently accepted value because of a flawed distance estimator. The history of the discovery, like so many, is complex and involved many others, including Hubble's assistant, Milton Humason. The value used in the scientific literature is 70 km/s per Mpc. This corresponds to 15 miles/sec per million light-years distance to somewhat less than 15% accuracy.

light-years across. We will start with the Milky Way in the center. If the galaxy named "Nan" was 25 million light-years away, it would be moving away at 375 miles per second according to the value of Hubble's constant ([15 miles/sec per million light-years] × [25 million light-years] = 375 miles per second). If another galaxy named "Orr" was 50 million light-years away, it would be moving away at 750 miles per second, and if "Pam" was 75 million light-years distant, it would be receding at 1125 miles per second. These are depicted in a row in the top panel of figure 1.3. Even with these enormous distances, the speeds are less than 1% the speed of light.

Now imagine that you could be instantly transported from the Milky Way, in the center of figure 1.3, to Nan. That is, you would be at rest on Nan. Of course, if you looked back at the Milky Way while sitting on Nan, it would be moving away from you at 375 miles per second. Here is a way to think about how the whole picture changes. If you were on the Milky Way and wanted to be at rest with respect to Nan, you would need to move at 375 miles per second to the right. This is shown in the middle frame. Moving next to something with the same speed it has is the same as being at rest with respect to it. This is just like looking at a car next to you on the highway going the same speed. Relative to you, that car is stationary. In the bottom illustration, we have just subtracted the velocities[9] to see what the universe looks like from the perspective of someone on Nan. But now note that the bottom picture looks just the same as the top but from the perspective of someone on Nan. Here again, the same Hubble-Lemaître law applies. Someone on Nan suspects they

9   A velocity is a speed with a direction associated with it. To subtract velocities, take the arrows in the middle, reverse their directions, and add them to the top row.

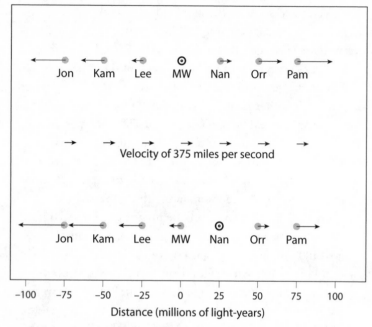

FIGURE 1.3. The expansion of the universe in one dimension. The top row shows the expansion from our point of view. The "MW" stands for the Milky Way. The circle indicates that it is reference point. The arrows indicate velocity. The galaxy Nan, at 25 million light-years' distance, is moving away from the Milky Way at 375 miles per second. Since Orr is twice as far away as Nan, it is moving away twice as fast as indicated by the velocity arrow being twice as long. The middle row shows the speed of Nan, but at all points in space, not just at Nan's location. Say you were moving at this velocity and near Nan. To you it would seem as though Nan was standing still. The bottom row shows the pattern of velocities from the point of view of someone standing on Nan. The circle shows Nan is now the reference. As you can see an observer on Nan seems to be the center of the universe and all the other galaxies are rushing away and obeying the same Hubble-Lemaître law.

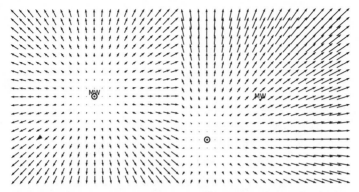

FIGURE 1.4. The expansion of the universe but now in two dimensions. Each point represents a galaxy. The arrows show the velocity the galaxy is moving as we see it. Of course, in reality the galaxies are much more irregularly spaced. The left side shows what we'd see looking out to greater distances than in the previous figure. The circle around the Milky Way indicates that it is the reference. Again, it looks like the galaxies are all moving away from us in proportion to their distance. Imagine that instead we transport ourselves to the galaxy marked by the thick arrow four galaxies in and six up from the lower left and so that we are at rest with respect to it. The right shows the velocities with this new galaxy as the reference. The overall picture is the same: we seem to be at the center with all other galaxies rushing away from us with speeds proportional to their distance.

might be at the center of the universe. For the time being, just imagine that this line of galaxies can go on forever and that the velocities increase without limit.

The main point here is that as long as the speed of recession is proportional to distance, all observers in the universe see the same pattern of recession and to all it *appears* that they are in the center of the expansion. Although we have shown the expansion with a line of galaxies in one dimension, it works in two and three dimensions as well. In figure 1.4 you can see the same process in two dimensions from the perspective of two widely separated galaxies.

THE BASICS

There is a simple and yet radically different way of thinking about the expansion. In the picture we just gave, we had in our minds a fixed space in which the galaxies were moving. That is, the *space* was fixed and the galaxies were traveling through it at different velocities. We now want to make a huge conceptual jump. Let us again imagine that the galaxies are on a line as in the top row of figure 1.3, but let's ignore the velocity arrows. Think of the galaxies as representing coordinates in space, just like mile markers on a highway (in one dimension) or as latitude and longitude positions on a two-dimensional map. Instead we want to add space between the mile markers. In the top row, this is equivalent to taking a pair of scissors, cutting the figure vertically between all galaxies, and taping an extra strip of paper of width, say, 0.2 cm wide. Let's say it takes us 30 minutes to complete this process. After we are done, Nan is 0.2 cm farther from the Milky Way than it initially was, Orr is 0.4 cm farther away than initially, and Pam is 0.6 cm more distant than initially. In this 30 minutes, Pam has moved three times as far as Nan, and Orr has moved twice as far as Nan. Pam, who started out three times farther than Nan, has moved three times as far in the same 30 minutes and thus *apparently* has three times the speed. We have reproduced the cosmic expansion but from a completely new perspective. Instead of space being fixed with the galaxies moving, the space between the galaxies is expanding.

From now on we want to think of space as fungible, not as a set stage on which the cosmic evolution unfolds but rather as the entity that is evolving.[10] In the above, Jon, Nan, and

---

10   For the experts, what we term "expanding space" is increasing the scale factor $a(t)$ in the metric. There are heated discussions about whether "expanding space" is a useful concept. In appendix A.2 we discuss some of the pitfalls.

Pam don't have to communicate with each other. They just sit still while space is created locally everywhere at the same time and at the same rate. In this picture, the Hubble-Lemaître law is just a statement about a specific rate of the expansion of space. In the two-dimensional case shown in figure 1.4, we cannot cut strips of paper but instead can imagine the galaxies as painted dots on a rubber sheet. Here, expanding space would be like stretching a very, very large rubber sheet in both dimensions. In three dimensions, we can again imagine an endless grid of Tinkertoys with the wooden hubs as the fixed coordinates and the struts between the hubs as the space that grows with time.

Although we have introduced a few analogies—making space by adding strips of paper, an expanding rubber sheet, Tinkertoys with growing links—we need to keep in mind that these are just analogies for describing the mathematical structure of general relativity. In space, there is nothing that acts like paper, rubber, or wood, but these different analogies are useful for different situations. The theory is more subtle and deep than we can convey with simple models and everyday objects.

To reiterate, though, we can think about the expansion of the universe as the expansion of space. The rate of expansion depends on the rate at which we "make space." We should *not* think about the expansion of the universe as galaxies flying away from each other in a pre-defined space. The Big Bang was *not* like a bomb exploding billions of years ago. The Big Bang marked the beginning of an explosion of space, everywhere at a fixed time in our distant past.

To recap, we started off this section by explaining the Hubble-Lemaître law and showed that no matter where you are in the universe, you appear to be in the center with the speed of recession of other galaxies proportional to their distances. We then introduced space as the changing quantity, and realized that if the galaxies represent fixed coordinates, the Hubble-Lemaître Law describes space expanding at a specific rate. In general we can expand space at any rate and still not be in a special place. We continue to think of the universe as infinite in extent.

Our new way of thinking about space begs the question "What is space?" This is a deep question, akin to "What is a vacuum?" Most physicists would say that we do not know. There are epochs in our cosmic history when an expanding space is the best description of Nature, and there are epochs when it can be misleading, leading us to imagine forces that do not exist. Regardless, an expanding space is a unifying concept that helps us envision the expansion of the universe and dovetails nicely with the warping of spacetime described by general relativity. We consider other elements of expanding space in appendix A.2.

Before moving on, we note that on human scales the expansion is ignorable. We see it only because we can look out to such vast distances. The forces that hold the Earth together and that bind the Earth to the Sun completely dominate the effects of an expanding space. Even our galaxy is not expanding. Gravity binds it together. We can be more quantitative. In 100 years, the width of this page would expand by about 0.001 microns, or roughly ten times the diameter of an atom if it partook in the cosmic expansion. This would be measurable. However, the forces that bind the molecules in the paper, which by measurement appear to be constant in time, would keep the page at its current size.

CHAPTER ONE

1.3 The Age of the Universe

If the galaxies are all apparently moving away from us now, they were closer together in the past. The universe used to be more compact: those galaxies in the Hubble Ultra Deep Field were ever closer together as we go back in time. To be specific, we will use the term "compact" to mean smaller in length or distance as opposed to volume. If the diameter of a sphere were halved, we would say it is twice as compact even though its volume would decrease by a factor of eight. When the universe was twice as compact, the objects in it were half as far away from each other.

At some point in the distant past, the galaxies were much, much closer together. If we go back even earlier, the galaxies had not yet formed, and instead of thinking of the space between galaxies, we think of the space between the constituents of galaxies. As we go farther back there was less and less space and, since there is the same amount of matter, the matter *density*[11] becomes enormous. At some point in our extrapolation the currently known laws of physics break down, but we need not extrapolate back that far. The important point is that we can extrapolate back to a time when the universe was extremely dense and that we reach that epoch in a fixed amount of time. In other words, the universe has a finite age.

The most accurate measure of the age of the universe comes from the Wilkinson Microwave Anisotropy Probe (WMAP) and the Planck satellites. The best estimate, taking

---

11 The matter density is the mass per volume. Similarly, we can have an energy density which is just the energy per volume. Cosmologists freely convert from one to the other using Einstein's celebrated relation, $E = mc^2$ where E is the energy, m is the mass, and c is the speed of light.

into account everything we know about the expansion history, gives 13.8 billion years to roughly 1% accuracy, that is the age is between 13.7 and 13.9 billion years.

We can get an approximate value of the age of the universe from the observations we explored in the previous section. As we saw, two galaxies 50 million light-years apart are moving apart at 750 miles per second. At this speed, assuming it is constant, 12.5 billion years ago the galaxies were on top of each other. Two galaxies 100 million light-years apart are now moving apart at 1500 miles per second, and so in the same amount of time, they too would be on top of each other,[12] as depicted graphically in figure 1.5. By extension all observers, no matter how far away they are from each other, would say the universe is 12.5 billion years old as all galaxies would be on top of each other when all the space is taken away. It is fortuitous that this simple estimate is so close to the more accurately determined value of 13.8 billion years, as we discuss next.

When we extrapolate back as we just did, we assume the rate of expansion has been constant. However, we know it has *not* been constant. At the very least, because gravity is a purely attractive force, the galaxies will be pulled toward one another, tending to slow the expansion. With just this simple observation we are linking the presence of matter to the rate of expansion, a concept at the heart of general relativity. Since the expansion rate has not been constant, the Hubble constant

---

12   A speed of 750 miles per second is the same as a distance of 4 million light-years per a time interval of one billion years. The age is then [50 million light-years]/ [4 million light-years/ 1 billion years] =12.5 billion years. Similarly, a speed of 1500 miles per second is the same as 8 million light-years per billion years. The age is then [100 million light-years]/[8 million light-years/ 1 billion years] =12.5 billion years. We give three significant digits so you can check the math. Using a more precise value of the Hubble constant would give 14 billion years.

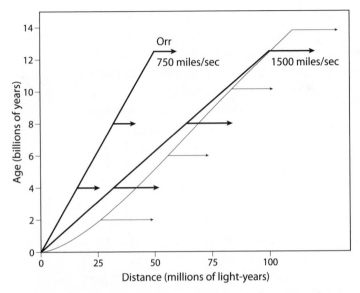

FIGURE 1.5. We imagine ourselves sitting on the *y-axis* at the ages marked there. Galaxy "Orr," as in figure 1.3, is at a distance of 50 million light-years and moving away at 750 miles/sec. Similarly, a galaxy at a distance of 100 million light-years is moving away at 1500 miles/sec. In the approximation that the speed of a galaxy does not change, the thick black lines show that as you go back in time, the galaxies get closer and closer together until they are on top of each other. As you can see, the Hubble "constant" is larger in the past even in this simple approximation. The thin gray line shows the true trajectory of a galaxy that was 100 million light-years away when the universe was 12.5 billion years old.

has not been the same throughout our cosmic history. The actual value depends on cosmic epoch and so it is often called the Hubble "parameter." The value we gave above, 15 miles/sec per million light-years, is only applicable now.

As recently as the 1990s there was no consensus on how to extrapolate back because we did not know the expansion rate throughout the history of the universe. However, as the web of observations has become ever tighter, we have learned about the cosmic composition and consequently the expansion history. We are now reasonably confident that through extrapolating back in time we know the age of the universe.

In some models of the universe, the current expansion is just one of many, perhaps one of an infinite number, of expansion cycles. As yet, there are no observations as yet to rule such models in or out. Most cosmologists do not subscribe to the cyclic model because it has not been as widely scrutinized as the baseline "inflation model" which we discuss later. However, we should keep in mind that cyclic models are a possibility. If in fact the universe is cyclic, then the 13.8 billion years would refer to the age of this cycle and the discussion in this book would refer only to the latest cycle.

We now have the framework to be more precise about the term Big Bang. We will take it to mean the time at which the universe began to expand. It is when we start our clocks. The term has nothing to do with space. We do not yet know enough physics to extrapolate all the way back to the Big Bang, even though many cosmologists are working on the problem.

In section 1.1, when we talked about the distance of an object, we imagined the universe as frozen in time. We can now appreciate the reason for this. There is a difference between the distance to an object when the light we observe was emitted and the distance at this instant after accounting for the expansion of the universe. At this instant, the object is farther away than when it emitted the light. Instead of talking about how far away something is, from now on we will mostly talk in terms of the compactness of the universe when the object emitted the light we observe. Equivalently, we can label the object by its age when it emitted the light we observe. By referring to compactness and age, we sidestep the fact that the universe expands while light travels to us. We say, then, that the most distant objects in the Hubble Ultra Deep Field emitted their light when the universe was about ten times more compact and 0.4-0.7 billion years old and don't mention distance. We follow the same practice for events and epochs as well as for objects. For example, about 5.9 billion years after the Big Bang, about 8 billion years ago, the universe was twice as compact. Cosmologists use slightly different terminology and say that the "scale factor" is 0.5 because distances between objects when the universe was twice as compact are half (0.5) the current size. For the most part, we'll use compactness, but there are times when the scale factor is easier. For example, the Earth and Moon formed about 9.3 billion years after the Big Bang, or roughly 4.5 billion years ago, at a scale factor of 0.71. Dinosaurs roamed the Earth 0.1 billion years ago when the scale factor was 0.993, and *Homo sapiens* appeared a mere 100,000 years ago when the universe was only negligibly more compact than it is today. Appendix A.3 provides a time line of significant events and the associated cosmic compactness.

In the previous two sections we added the element of time to the static picture we started with in section 1.1. Let's put this in the context of Tinkertoys. We now see that 13.8 billion years ago, all the wooden hubs were on top of each other and smushed together. Since, as far as we are able to tell, the Tinkertoy grid is endless in space today, it is still endless after rolling back the expansion; but, as we do so, the links get increasingly shorter and thus the whole grid structure becomes much more dense. We cannot extrapolate back to infinite density or zero time: we don't know how because the laws of physics break down. Although we have learned a lot, the picture is not fully satisfying because we still haven't explained why we can count all the objects in the observable universe in the direction of the Hubble Ultra Deep Field. We tackle that next.

1.4 The Observable Universe

We just saw that the age of the universe is finite and that all observers agree on its value. The next key ingredient for developing our model is taking into account the speed of light. So far, we have primarily used the finite speed of light to establish a distance, namely the light-year. We continue in that vein.

If we could instantaneously travel anywhere in the universe *right now*, the galactic environment would look similar to the environment around us. This is the cosmological principle. There would be a variety of galaxies, but no matter where we went we would compute the age of the universe to be 13.8 billion years.

Because the speed of light is finite and we know the age of the universe, there is an upper limit on the size of the universe we can observe. In other words, the size of the observable universe is finite. It is easy to get an estimate of this size. To a first approximation, in one direction we cannot see farther away than the age of the universe times the speed of light. It is as though each observer is in the middle of a spherical volume with a diameter of $2 \times 13.8 = 27.6$ billion light-years. The actual diameter of the spherical volume is a little more than three times larger because our approximation did not include the fact that the universe expands while light is traversing it.

Nevertheless, the important conceptual point is that because information cannot travel faster than the speed of light there is a limit to how far we can see, hence the name the "observable universe." When cosmologists talk about the "universe," often they really mean the observable universe. It is good to keep in mind, though, that at this instant, the galactic environment at the "edge" of our observable universe is similar to what we see around us. See appendix A.4 if you'd like additional details on the relation between the age, size, and compactness of the observable universe.

1.5 The Universe Is Infinite?

Far, far beyond our observable universe, space—and even the laws of physics—might be different. We do not know whether the universe is truly infinite, in the sense that it goes on forever in space. However, observations tell us that an infinite universe with properties similar to the cosmic environment around the Milky Way is the best and most parsimonious description of the data. That is, we cannot tell the difference between what we observe and a model of the universe that is infinite in spatial extent.

To put this in context, until a few decades ago there was no scientific reason to believe a priori that the universe was infinite. Cosmologists did not known whether continued observations would tell us that the universe was finite, in other words that it had a finite extent and contained a fixed amount of stuff. If so, we would still have an expanding universe with a finite age, but it would be finite in extent and would collapse in a finite time. Instead, the observations of the contents of the universe, which we describe in the next chapter, have told us that for all intents and purposes we should treat the universe as infinite.

For a moment, let's picture the universe as an absolutely enormous container of chocolate chip ice cream. Think of the chocolate chips as the galaxies and the ice cream as the space. Our observable universe would be like a very large scoop taken from somewhere inside the container, far away from the container walls. Our scoop would have all sorts of different sized chips. But all scoops would be similar and recognized as the same chocolate chip ice cream no matter where we took them, as long as they were well away from the walls. The container walls, if they exist, represent some new physics to which we have no access.

The extent of the universe is an active area of investigation. Every now and then someone comes up with a model for a finite universe. However, when the predictions of the model are compared to the data we find that an infinite universe provides a better description of the observations. With this picture in mind, the question "What is the universe expanding into?" is not answerable or even relevant.

1.6 How to Look Back in Time

We now add the next conceptual component to our picture. This one is again based on the speed of light, but here we are not using light as a measure of distance. Earlier we noted that since the Sun is eight light-minutes away, we see it as it was eight minutes ago. Similarly, if an object is 20 million light-years away, when we observe it we see it as it was 20 million years ago. As we peer deeper and deeper into space we see objects as they were at earlier and earlier stages in their lives.

Our whole cosmic history can be read by looking ever deeper into space, because as we do so we look farther back in time. In other words, telescopes are like time machines.

First let's think about what this means taking just a small step out. Stars can explode, releasing enormous amounts of light and particles in a "supernovae." We are able to see these explosions. In 1987, we saw that a star exploded in the neighboring Large Magellanic Cloud (plates 3 and figure 1.2). The Large Magellanic Cloud is about 160,000 light-years away. That is, the star actually exploded before *Homo sapiens* first came on the scene, but we only saw the light in 1987. This particular supernovae, called 1987A, is unique because in addition to the light, we detected neutrinos from it. Neutrinos are elusive elementary particles associated with nuclear interactions. They can travel at close to the speed of light and they barely interact with matter. We will return to them in more detail later, but for our purposes here, note that it is not just light that comes to us from the distant past but particles as well. From this supernova alone, roughly one hundred billion neutrinos per square centimeter hit the Earth. Most passed right through. Twenty-five were detected by the Kamiokande detector in Japan.

Supernovae are so bright they can be seen to vast distances. With powerful telescopes, astronomers can catch supernovae from relatively short-lived stars that exploded when the universe was twice as compact as it is now, 5.9 billion years after the Big Bang. That means the original star has not existed for 8 billion years! What is left for us to detect is a roughly spherical shell of light and particles traversing the universe. We see this shell as it passes by Earth and as the billions of particles associated with it stream through us as if we didn't exist. Similar supernovae are taking place all over the universe and sending out their blast waves to travel the cosmos. For the ones we detect, we can study the dying embers of the explosion to understand the composition of these distant stars.

Although individual young stars at great distances are too small to see unless they explode, we can see nascent galaxies from when the universe was less than a billion years old. Let's go back to the Hubble Ultra Deep Field. Plate 5 shows us what we see as we peer deeper and deeper into space. Through a combination of sensitivity and ability to focus intensely on a small patch of sky, the Hubble and other telescopes can almost look back to a time when galaxies were just forming. Earlier we said that we could count all the galaxies in the universe. Now we can see what this means. We can look back to a time before galaxies existed, which corresponds to an epoch when the universe was about 20 times more compact and about 200 million years old (see appendix A.3). Thus, in the part of the universe to which we have access, our observable universe, we can count *all* the galaxies. Again, there are about 100 billion broadly similar to the Milky Way.

If we peer deeper still we could see the birth of the first stars. This has not yet been done, but instruments are being built to make this possible. Going back even farther in time, we can see the remnant radiation of the Big Bang, the Cosmic Microwave Background. It is the light from the edge of the observable universe.

We have presented a vast and expanding framework with which to think about the universe. The expansion forces us to think about a more compact universe in the past. We extrapolated the expansion all the way back to a dense Big Bang that took place 13.8 billion years ago. This is the age of the universe. The finite speed of light combined with this fixed time led us to realize that we can only look so far out into the universe. In other words, we can access only the observable universe.

For much of the preceding discussion, the galaxies were just distance markers or signposts that helped us understand ideas such as "the observable universe" and that our universe has a definite age. We could have developed the same picture with just a small fraction of the galaxies we actually observe. We were, after all, only considering aspects of space and time as linked by the speed of light. But, as we will see, the contents of the universe and the expansion of space are intimately related. As a first step of making the connection, let us turn to the question of what the universe is made.

CHAPTER TWO

THE COMPOSITION
AND EVOLUTION
OF THE COSMOS

THERE ARE THREE MAJOR COMPONENTS TO THE UNIVERSE: radiation, matter, and dark energy. We think of each as a density—that is, as an energy or mass per volume. As mentioned earlier, by using $E = mc^2$, we can convert a mass to an energy, or vice versa, and treat the three components on the same footing. We can then say that the energy density of the cosmos, averaged over a large volume, is made up of x% radiation, y% matter, and z% dark energy. Let's briefly review these three terms before getting into the details of the x, y, and z.

The universe is filled with radiation in the form of thermal energy. This is the CMB. As we shall see, the CMB really is a fossil of the infant universe, but it is a fossil in primordial light as opposed to something more tangible. Like a dinosaur footprint, it is not so important for understanding the current state of the universe but it is essential for telling us how we got to where we are.

The matter component is divided into two subcomponents, atoms and dark matter. When we peer deep into the night sky, with say the Hubble Space Telescope, we see galaxies because the atoms in them emit light. Not only are the atoms we see a small fraction of all the atoms, but the atoms taken together are just 17% of all the mass, and furthermore all the mass accounts for just 30% of the total energy density. When we observe an image of galaxies, it is as though we are flying over land at night and trying to figure out what's beneath us—mountains, forests, deserts, lakes—by looking at the distribution of house lights. The house lights are like galaxies and Earth's surface is like the universe. In areas near cities you can tell what's below you, but for most of the flight you need more than just a snapshot of the lights. By observing the universe in different ways, we can get that additional information to determine the cosmic composition.

The third major component is the dark energy. In contrast to the CMB, it *is* important for understanding the current state of the universe and its future expansion, but was insignificant in the early universe. It is the component we understand least. We've only known of its existence since the late1990s and are still trying to connect it to the rest of physics.

At different epochs in our cosmic history, one of these three forms of energy density dominates over the others. For the first roughly 50,000 years of cosmic history, radiation in the form of the CMB was the dominant form of energy. Then, for the next ten billion years, matter dominated; or, to put it in the same terms as the CMB, its equivalent energy dominated. And, most recently, for the past 3.8 billion years, dark energy has come to dominate. We now examine these three components in more detail to see how they interact over time to produce cosmic structure.

COMPOSITION AND EVOLUTION

2.1 The Cosmic Microwave Background

The primary characteristic of the CMB is its temperature, 2.725 K. In this section we interpret what this means. Another aspect of the CMB is its small temperature differences from place to place in the universe, or, as viewed by us, from position to position on the night sky. A third aspect is its polarization. We will address temperature differences and polarization in later sections.

The fact that we can characterize the CMB as a temperature is a profound statement on its own. The CMB is thermal radiation, or radiant energy, of a very particular form called "blackbody radiation." Things that emit blackbody radiation are called blackbodies.

To get a sense for thermal emission, let's consider a simple comparison. A black piece of paper left in the Sun gets hotter than a white one, which gets hotter still than a perfect mirror. The black piece of paper absorbs the radiation that lands on it, the white piece of paper absorbs some of the radiation but scatters most of it away, and the perfect mirror reflects all the radiation that lands on it and doesn't absorb any.[1] From the laws of thermodynamics, we can deduce that a good absorber of radiation is also a good emitter. So, if you put your hand over, not on, the black piece of paper that has been exposed to the Sun, you will feel that it radiates more energy than the white one or a mirror. Even better examples of blackbody radiators are the Sun or a pottery kiln.

Objects emit their thermal energy over a range, or spectrum, of wavelengths. However, even for blackbodies, most

---

1 It is difficult to make a perfect mirror. Aluminum is an obvious candidate, but it is a good absorber of ultraviolet radiation and so gets hot in the Sun. If our eyes could see ultraviolet radiation, which they cannot, an aluminum mirror would look dark.

of that energy comes out over a limited portion of the spectrum; in other words, blackbodies emit predominantly over a relatively small span of wavelengths. For the Sun, almost half of the energy comes out at wavelengths between 0.4 and 0.8 microns. It is no coincidence that this is the visible spectrum we detect with our eyes, which likely evolved to take advantage of the Sun's spectrum. We know the Sun also emits UV radiation. For example, "UVB," the primary source of sunburn, is at 0.3 microns, but we cannot see that radiation. The Sun also emits in the "near-infrared" region, but we can't see that light either.

The lower the temperature of an object, the longer the predominant wavelength of emission. This is known as the Wien displacement law, which says that the dominant wavelength of emission of a blackbody in microns is roughly 3000 divided by its temperature in kelvin. For example, the Sun with its temperature of about 6000 K (see note 1 in the preface) emits predominantly at a wavelength of 3000/6000 or 0.5 microns. The Milky Way is about two hundred times colder than the Sun, or 30 K, so it emits predominantly at a wavelength that is two hundred times longer. Using our simple law, we find the wavelength is 3000/30 or 100 microns. This is the wavelength for the far-infrared radiation that DIRBE measured, as shown in plate 3. Although the Wien law applies to a single wavelength, we should really think of most of the radiation as coming from a range of wavelengths around the predominant one. With this simple relationship, we can link the temperature of an object to the wavelength of light it emits.

We can also think of thermal emitters such as the Sun in terms of atomic processes. The hotter an object is, the more the atomic constituents jostle around and emit light. The more they jostle, the more energy they emit; the more energy they emit, the shorter the predominant wavelength. The difference between a blackbody emitter and an ensemble of energetic atoms is that to get blackbody radiation you need a lot of atoms each absorbing the radiation of its neighbors and then re-radiating it. We might imagine the radiation as beach balls and the atoms as a special group of beachgoers who always play with beach balls. On a hot day there would be so many beachgoers that you'd need smaller balls just to play. From a distance you'd see a swarm of small beach balls energetically passed around and in the air. This corresponds to short-wavelength high-temperature radiation. On a cold day, there would be fewer people out so they could use larger balls, and we can imagine folks being less excited about playing the game, so there would be fewer balls in the air. This corresponds to long-wavelength low-temperature radiation.

It is a deep aspect of blackbody radiation that all you have to do is specify its temperature and you know how much energy is radiated at all wavelengths. That is, the temperature describes the entire spectrum, not just the predominant wavelength. By definition, objects that emit blackbody radiation are in thermal equilibrium with that radiation. In other words, the temperature of the radiation corresponds to the temperature of the object. Say you embedded a thermometer in the walls of a kiln well away from the radiation. The temperature that you would ascribe to the radiation in the kiln by measuring the amount of energy at each wavelength would be the same as the kiln's wall temperature as read by your thermometer. Using our earlier analogy, you can tell how many beachgoers there are, and the beachgoers' temperature, simply by looking at the beach balls in the air.

In 1900, Max Planck derived the celebrated formula that describes blackbody radiation. The CMB has by now been

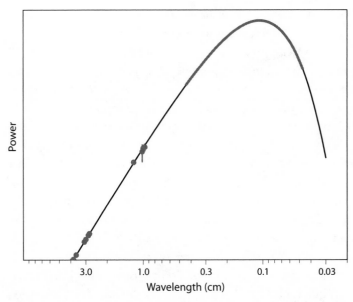

FIGURE 2.1. The spectrum of the CMB. The *x*-axis shows the wavelength and the *y*-axis shows the emitted power. The thin black line shows Planck's celebrated formula for a 2.725 K blackbody. The continuous gray line shows the measurement from the FIRAS instrument on the COBE satellite. The error bars are smaller than the thickness of the line. Some selected measurements at wavelengths longer than those from FIRAS are shown in gray as well. The agreement between the observations and the blackbody formula is clear.

measured at many wavelengths, and to the limits of measurement it follows the Planck formula. So, we know that it came from an era when the matter in the universe was in thermal equilibrium with the radiation. Figure 2.1 shows the measurement of the CMB spectrum from the COBE satellite and other instruments. Many have tried to explain the spectrum with different sources of radiation in order to find alternative explanations to the Big Bang. One proposal was that the CMB was emission by distant clouds of cool dust.

Such attempts have not succeeded because the predicted spectrum from alternative sources of radiation does not match the observations. Nevertheless, searching for departures from a Planck spectrum is important. A departure can tell us, for example, if there was an injection of energy into the universe, say from a decaying particle, from some earlier epoch.

In an historic step for physics marking the birth of quantum mechanics, Planck hypothesized that electromagnetic radiation was quantized to derive his formula. This means that radiation can be described as discrete packets or quanta of energy. These quanta are called "photons" or "particles of light." Part of the foundation of quantum physics is that the interaction of radiation and matter may be considered either as involving waves and matter or as involving photons and matter. At times one formulation is easier to work with than the other. Our beachgoers playing with beach balls are analogous to atoms absorbing and emitting photons. For the CMB, there are currently 400 photons in every cubic centimeter of the universe. Once we know the radiation is a blackbody, specifying the photon density is equivalent to specifying the temperature.

We do not know a priori* that the universe started off in an incredibly hot state. Leaving the CMB aside, the whole picture we have been developing of the expanding universe could in principle work with a relatively cool early universe. However, because the CMB exists, we know the early universe was hot and in thermal equilibrium. Here is how it works.

When the universe expands, the wavelengths of light are stretched in proportion to the expansion. Imagine you had a slinky. Think of each full turn of your slinky as corresponding to a wavelength of light. Let's say that the slinky is initially

10 cm long. Now stretch it to 20 cm. The total number of turns is the same but the space occupied by each turn has increased. This is analogous to the stretching of the wavelength of light as the universe expands by a factor of two.

Not only do the wavelengths of the CMB get stretched on their way to us, but the wavelengths of all light from all distant objects get stretched. We see distant objects not only as they were when they were younger, we also see them through stretched wavelengths.

There is another way to think about the stretching wavelengths. A state trooper patrolling the highway might point a Doppler radar gun at your car to see how fast you're driving. When the radar beam bounces off your car and back to the trooper, its wavelength is slightly shifted, in fact shortened if you are moving toward each other. This is known as the Doppler effect. It occurs because through the reflection, your car in effect becomes a moving radar source; and a moving source emits a different wavelength than one at rest. The trooper can tell how fast you are moving from the difference in the transmitted and received wavelengths. The shift is small but may be computed accurately with the Doppler equation. If instead the trooper receives a signal from a source moving away, its wavelength is stretched. Because red is at the long wavelength end of the visible spectrum, we say the light is redshifted.

Hubble and Lemaître used the redshifted light from recognizable atoms in distant galaxies to determine their speeds. There is an interesting subtlety, though, related to our earlier discussion of how to think about the expanding universe. For very distant objects, much farther away than those known to Hubble and Lemaître, that are moving away at an appreciable fraction of the speed of light, the recessional speed is not described by the Doppler equation or even its relativistic extension. The apparent speed of a distant object, the one that enters in the Hubble-Lemaître law, is from the expansion of space. This phenomenon is called the cosmological redshift. Let's now apply these concepts to the most distant light we observe, the CMB.

If we go back in time, the wavelengths that comprise the CMB decrease because space is more compact. At its current temperature of 2.725 K, the most prominent wavelength of emission is 3000/2.725 or approximately 1000 microns (0.1 cm as in figure 2.1) as determined by the Wien displacement law. You can show that the Planck formula keeps its same form if the universe is more compact, but to do so the temperature has to increase inversely with its size. Going back in time to when the universe was twice as compact, as it was 8 billion years ago, the temperature of the CMB was double the current temperature, or about 5.2 K: its predominant wavelength was half of what it is today, or 500 microns; there were 3200 photons per cubic centimeter; and the spectrum was still that of a blackbody.

We can keep going back to when the universe was much more compact. At 400,000 years after the Big Bang when the universe was 1,000 times more compact, the CMB was 2725 K, about half as hot as the Sun. It was just about to be energetic enough to rip electrons away from the proton nuclei in hydrogen atoms.

Going back even more, to about three minutes after the Big Bang, when the universe was about a third of a billion times as compact as today and thus at a billion kelvin, the radiation was so intense that the nuclei of helium could just barely hold together. The energy that binds the neutrons and protons together in a helium nucleus is about a million times greater than the energy that binds the electrons to the protons, so the radiation has to be about a million times hotter, with wavelengths a million times shorter, to rip the nuclei apart.

At yet another factor of 3,000 times more compact and hotter, about 25 millionths of a second after the Big Bang, neutrons and protons did not independently exist and the universe was a "quark-gluon plasma." (Quarks are the elementary particles that make up protons and neutrons.) This state of matter has been reproduced on Earth in the Relativistic Heavy Ion Collider (RHIC) on Long Island, New York. If we go back to a state where the universe was about 50 million billion times as compact as today, about a hundred thousandth of a billionth of a second after the Big Bang, the energy in the photons was roughly the energy of a collision between protons in the Large Hadron Collider in Geneva, Switzerland. These are the highest energy elementary particles produced by humankind so far. Yet, using the universe we can explore even greater energies.

Let us now tie in the idea of a hotter younger universe with the spatial picture we have been developing. If we could travel anywhere in the universe *instantaneously* right now, the temperature would be 2.725 K everywhere. We can call this the current temperature of the universe. If instead you could have traveled *instantaneously* around the universe when it was twice as compact, you would have measured the temperature everywhere in space to be 5.2 K. A galaxy at this time would be 8 billion years younger. If we observed this same galaxy today from Earth, we would see it as much younger *and* the wavelengths would all be a factor of two longer because of the expansion of the universe since the time the galaxy emitted its light.

COMPOSITION AND EVOLUTION

When we measure the CMB, where does the light come from? Let's go back to plate 5. The CMB light that lands on our detectors has been traveling to us since just after the Big Bang. It started on its path toward us before there were stars or galaxies and of course before the Earth existed. Back then, it was more energetic and was still described by the Planck function, but with a much, much higher temperature. On its way to us, the universe expanded, the wavelengths stretched, and the radiation cooled. We now see the remnant glow of the Big Bang 13.8 billion years ago in a conceptually similar way to how we see the supernovae light from stars that no longer exist. Unlike the supernovae, the CMB comes to us from all directions.

To summarize, early in our cosmic history, when the CMB was incredibly hot, it was the dominant form of energy density. Being in the universe then was like being inside an unimaginably hot and large pottery kiln. Now, because of the expansion, the CMB is but a cool, dim afterglow with a nearly negligible effect on the current universe. As the universe expanded and the CMB dimmed, matter became the dominant form of energy density, leading to a new set of phenomena. Most important, it allowed structure to form. Before fitting the pieces together, we need to explain a little more about matter.

2.2   Matter and Dark Matter

All matter with which we have direct experience is made up of protons, neutrons, and electrons. These are the building blocks of atoms. The atoms interact with each other both gravitationally and through the exchange of photons, that is through various wavelengths of light. To be sure, there are other fundamental particles and other forces, but mostly they are not part of our daily lives.

Before addressing dark matter, let's review the known particles and atoms most relevant to cosmology. The simplest atom is hydrogen. It consists of one positively charged proton in the nucleus orbited by a negatively charged electron roughly one ten-thousandth of a micron away. The proton is 2,000 times heavier than the electron. If the hydrogen is in a high-temperature environment, the electron can be ripped off by the energetic photons so that both the proton and electron are free. The hydrogen is then said to be ionized. The next simplest atom is deuterium, which still has only one electron but has a proton and neutron in the nucleus. A neutron has roughly the same mass as a proton but is neutral. Because deuterium has the same number of protons, it is called an isotope of hydrogen, and is often referred to as "heavy hydrogen."[2] Going up in mass, the next familiar atom, and the next one in the periodic table, is helium. It has two protons and two neutrons in the nucleus orbited by two electrons.

All the other elements are formed from these basic elements, as we will describe later. When we peer out into the night sky with telescopes, we see that the light from distant galaxies comes from some of the same atoms we find on Earth; not just the simple ones mentioned in the previous paragraph, but a host of more complex atoms and molecules as well. These distant galaxies are made of the same stuff we are. This simple observation suggests a common origin.

---

2 When the hydrogen in water is replaced by deuterium, you get "heavy water," which is used in some nuclear power plants.

COMPOSITION AND EVOLUTION

There is one more particularly relevant fundamental particle in cosmology—the neutrino. As its name implies, like the neutron, it is neutral. Neutrinos barely interact with anything. They are products of nuclear decays and nuclear interactions. For example, a free neutron will decay, on average, in just over ten minutes to a proton, an electron, and a neutrino.[3] As another example, the fusion reactions that power the Sun (and sustain life on Earth) generate about 100 billion neutrinos that go through each of our fingernails each second. We are sieves to these particles. They go right through the Earth as well.

Because they interact so little, neutrinos are especially difficult to study. We know there are three types, but we don't know the masses of any of them. All we know is the mass difference between different types. Multiple experiments are being conducted around the world to determine their properties. From the nuclear interactions in the early universe, there should be 300 neutrinos per cubic centimeter throughout the universe today traveling at a few percent the speed of light. This means roughly the same number of neutrinos from the early universe go through each of your fingernails in one second, as do neutrinos from the Sun. Even though there are almost as many primordial neutrinos as CMB photons, they have yet to be detected.

The early universe is simple. From about the time that protons and neutrons emerged from the quark-gluon plasma until the first stars formed some 200 million years after the Big Bang, the most important forms of *known* matter are the proton, neutron, electron, and neutrino along with their anti-

---

3 More specifically, the neutron decays to a proton, electron, and an electron antineutrino. To the limits of measurement, free protons do not decay.

CHAPTER TWO

particles. With regard to the matter in the universe, cosmic evolution is determined by how these four particles interact with each other, the CMB photons, and the gravitational attraction of the dark matter in an ever-cooling universe.

Dark Matter

If you looked up in the night sky and saw that, over a period of time, a distant star was following a circular path of, say, two full moons (a degree) in diameter, you would immediately conclude that it was in orbit around another object. For an object to move in a circle there *must* be a force acting on it.[4] In the cosmos, that force is gravity. You might then train your telescope to look for the companion. You know that something has to be applying a gravitational force on the star. There has to be some "missing matter." It might be a black hole or a dim star that you had not noticed at first.

For many decades, astronomers have been making observations similar in spirit to the one above (although in much more clever ways) with different systems and with less obvious geometries. In the cosmological context, the existence of missing matter was first proposed by Fritz Zwicky in 1933 based on observations of the Coma cluster of galaxies. Others extended his findings. Of particular note were observations of the orbital velocities of stars and star-forming regions as well as of diffuse hydrogen gas in the Andromeda galaxy, an excellent laboratory because, as shown in figure 1.2, it is nearby and looms large. In 1970, Vera Rubin and Kent Ford showed clearly that the velocities of the stars they observed agreed with earlier measurements of the diffuse hydrogen gas velocity. Subsequent models of the orbits of the observed stars and

---

4 This follows from Newton's second law.

gas in Andromeda showed that there had to be additional matter that was not in either luminous stars or diffuse gas in order to explain the measured velocity profiles. More generally, regardless of the size of the system, from nearby in our galaxy to distant galaxies and groups of galaxies, astronomers found that there is not enough observable matter to account for the motions of stars and galaxies.

The amount of missing matter is not small, nor is its effect subtle. By observation, there is more than about five times as much missing matter as observable matter. The best accounting of it comes from measurements of the spatial variations in the CMB, which we discuss in chapter 3. Here, though, we focus on the characteristics of the missing matter that are independent of the CMB.

The path from not being able to find the missing matter to concluding that there must be a new form of invisible matter, or dark matter, has involved thousands of scientists and multiple lines of evidence. In part the missing matter has been characterized by process of elimination. We know what it is not. We know it cannot be an assemblage of planets, say "Jupiters," that are just too dim to see. We know it cannot be atomic, such as hydrogen gas, or be the same as the stuff of which we are made. We know that it cannot be black holes of the type that have been observed so far. We know that it cannot be one of the three types of neutrinos, even though there are almost as many neutrinos in the universe as CMB photons.

One assumption is that the missing matter is a new type of elementary particle, but it may just as well be a new family of particles, multiple families, or a combination of different types of particles. Generically we call these possibilities "dark matter." If dark matter is a particle, we do not know how it interacts with other particles or even, if two dark matter particles collide, with itself. We know that it does not interact significantly with photons, which is why it is called "dark." Observationally, all we know is that the dark matter interacts gravitationally. Its character is a grand mystery, it is unambiguous that dark matter exists in vast quantities and that it is not a form of matter we have encountered in our laboratories.

One of the clearest astronomical observations that shows the need for dark matter is of the Bullet Cluster, as seen in plate 6. The image actually shows two clusters of galaxies that have collided and passed through each other. The one on the right, as suggested by the pink shape, is the "bullet." Before they collided, the clusters were full of a moderately uniform mixture of normal matter in the form of a diffuse hot gas and stars in individual galaxies, and dark matter. In both, the amount of mass in the hot gas was much greater than in all the stars that made up the galaxies, and the mass in dark matter was much greater again than the mass in the hot gas in this system. When the clusters collided, the galaxies and dark matter passed through each other almost unscathed, but the gas interacted. Think of it this way. If you filled both hands with pebbles and threw them toward each other with their trajectories crossing a short distance in front of you, most of the pebbles would not collide. The pebbles are like the galaxies and dark matter. If instead you aimed two garden hoses at the same meeting point, the water from each hose would collide and interact. The water is more analogous to the hot gas.

The gas in the clusters is roughly ten million kelvin. It is so hot that it emits X-rays which are observed by NASA's Chandra X-ray Observatory. In plate 6 the hot gas is pink. In other words, the pink shows us the location of most of the normal matter. The blue shows us the location of primarily the dark matter, but also the galaxies. To understand how we

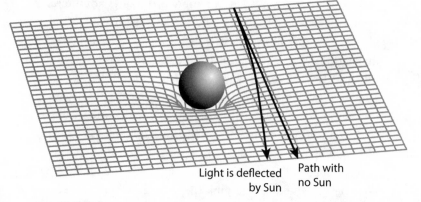

Light is deflected
by Sun

Path with
no Sun

FIGURE 2.2. An example of bending light in a curved two-dimensional space. The Sun is represented as the ball in the center of the image. It warps space as would a bowling ball on a large rubber sheet. Light's path by the Sun is analogous to a quickly rolled small marble going past the bowling ball. The marble follows the contour of the two-dimensional rubber sheet and its trajectory is bent toward the bowling ball, away from a straight path. It rolls along the easiest path. Similarly, in three dimensions a light ray going by the Sun follows the easiest path. Its trajectory is bent by the curvature of three-dimensional space or, equivalently, by the force of gravity. In the figure the deflection is greatly exaggerated.

know the matter is there we need to take a small detour to discuss the bending of light.

Let's go back to thinking about space. Not only can it expand but it can also be warped or curved. When light from a distant star travels to us on a path that goes close to the Sun, it is deflected ever so slightly. This can be thought of as the Sun's gravitational pull on the light. A better way to think about it is that the space around the Sun is curved and that the light from the distant star follows the easiest path on its way to us.[5] Figure 2.2 shows one way to visualize this. The bending

5 In general relativity, the description of the space around a massive object is mathematically different from the description of the geometry of the universe. For light

of light rays by the gravitational field of a large mass is directly analogous to the bending of light by a camera lens. The phenomenon is called gravitational lensing. You can determine the amount of mass from the amount of bending.

We can now understand the blue areas in plate 6. Distant galaxies far behind the Bullet Cluster were observed through the Bullet Cluster. From the distortion of the distant galaxies, the effective lensing and mass distribution in the Bullet Cluster were determined. The image shows that most of the mass is in two distinct regions. The essential feature of the image is that the dark matter is clearly separated from the normal matter. The hot diffuse gas interacted during the collision and got left behind.

The search for dark matter is a very active area of physics. Multiple experiments are trying to detect it directly. Some are trying to identify a direct hit of a dark matter particle with a target atom of germanium, argon, or xenon. These experiments are often constructed deep underground to shield the target atoms from other known particles that can't easily penetrate the Earth. Other experiments take much different approaches to search for different types of interactions and different forms of dark matter. There have been hints of possible detections and reported detections that have not withstood further scrutiny. As of 2019, no iron-clad direct detections have been made. We hope that dark matter particles will be detected in the Large Hadron Collider.

The discovery of new elementary particles has mostly taken place in particle accelerators that were precursors to the Large Hadron Collider. There is an enormously successful "standard

---

moving by an object, the effect of gravity on the passage of time is also a significant effect.

COMPOSITION AND EVOLUTION

model of particle physics" that has 17 different fundamental elementary particles, including the quarks that make up the protons and neutrons, the electrons and neutrinos, and most recently the Higgs boson. Although comprehensive, predictive, and well tested, we know the standard model of particle physics is not complete because there are measurements of elementary particles that it cannot explain, such as the mass of the neutrino. We hope that the detection and characterization of the dark matter in the lab will show us how to advance our model of particle physics.

Is it possible that there are no dark matter particles and that our laws of physics are incomplete? A lot of research has gone into investigating how general relativity might be wrong on large scales and so, in fact, there is no missing matter and instead a new force accounts for the observations. These new theories generally go under the name MOdified Newtonian Dynamics, or MOND. Fortunately, MOND makes predictions that can be tested, and some of the predictions do not agree with observations. In contrast, there has yet to be an observation in disagreement with general relativity. Therefore the large majority of cosmologists do not agree with MOND. Of course it is quite possible that there are other forces or laws of physics we simply have not yet discovered.

2.3 The Cosmological Constant

Earlier, we gave an approximate value for the current expansion rate of the universe as 15 miles per second for every million light-years' distance. Put another way, a galaxy 10 million light-years distant appears to be moving away from us at a speed of 150 miles per second. In the late 1990s it was discovered that the expansion *rate* is increasing. In other words, the expansion is accelerating. In one billion years that same galaxy will move away at 156 miles per second; one billion years ago it was moving away at 144 miles per second.[6]

This remarkable observation was made by two independent groups, the Supernovae Cosmology Project and the High-Z Supernovae Search Team, and has been confirmed by others. As their names suggest, they used supernovae to look back to when the universe was just a few billion years old. The trick was to find objects for which they could determine accurate distances and speeds and compare the expansion rate then to the current expansion rate.

One way to think about this observation is that space is being made at an accelerated pace. Not only is the concept of "expanding space" everywhere a convenient way of thinking about the expansion of the universe, but we are now almost forced to think about it this way. In a static space, we can imagine that two galaxies could be moving apart at a nearly constant speed, slightly slowing down due to their gravitational attraction, but we can't come up with a way for them to accelerate away from each other. Acceleration requires a force, and in a static space the only force available is gravity which, if anything, would tend to decelerate the expansion.

So the question is, why is space being made at an accelerated rate? We do not know. What it means is that space, the vacuum, appears to have an energy density associated with it. This energy density acts like a pressure that expands the universe or, more prosaically, "expands space." The energy density is quantified as a cosmological constant denoted by

6 An accelerating universe corresponds to the Hubble parameter approaching a constant value. This is because the Hubble parameter is the expansion rate divided by the scale factor (as given in appendix A.3).

COMPOSITION AND EVOLUTION

the Greek letter Lambda, Λ. This is a new constant of Nature that actually may not even be constant.

Einstein introduced Λ in 1917, before Hubble's observations. He thought the universe was static—that is, not expanding as Hubble's observations showed. To understand his motivation, imagine two isolated galaxies in the universe. They are attracted to each other by gravity and would fall toward each other. The cosmological constant provides a new kind of pressure that balances out the gravitational attraction between them and holds them in place. By extension, this would apply to a universe full of galaxies. However, after Hubble's observation, Einstein abandoned Λ. We now know this pressure exists at an even larger value than Einstein thought necessary.

There are other explanations for the accelerated expansion then the cosmological constant. In general they posit some form of "dark energy" that is not constant. These alternatives make predictions for the acceleration versus the age of the universe. Measurements are in progress to test these predictions. We do not know if the dark energy is, say, a substance, or if it is constant throughout space. Perhaps we are missing some fundamental element of one of our theories. At the moment, the most straightforward explanation that agrees with all the data is that space is described by a cosmological constant that is constant in space and over time. So, let's adopt this point of view.

The mere existence of the cosmological constant is deep. It is not part of any fundamental theory in physics. It has no bearing on life or physics on Earth. No laboratory experiments have been developed that can measure it. It is a constant that allows us to quantify the cosmic acceleration and in so doing tells us that there is an energy density or pressure associated with space.

CHAPTER TWO

What does this mean for the future? We will set aside the cautionary words in the introduction and extrapolate. If you are in your car on the highway and accelerate at a constant rate, you of course go faster and faster. The case is similar for the universe, but more extreme than a constant highway acceleration. In the universe, the space between galaxies is growing exponentially. Galaxies that are widely separated now will soon *apparently* be moving apart faster than the speed of light. Here is an example where some of the analogies we sometimes use to describe the expanding universe break down. It is not physically possible for a rubber sheet to expand in this way. There are simply no material objects that can act like the expanding universe.

There is no contradiction between the accelerating expansion and the special theory of relativity, which requires only that information and massive particles cannot be transmitted from one place to another faster than the speed of light. For the galaxies, the space in their cosmic neighborhoods is just expanding at an exponentially increasing rate. No information is being transmitted faster than light. From the point of view of someone in the Milky Way, distant galaxies that we can currently observe will simply fade away in the future. We do not know how long the exponential expansion will last.

Let's take account of where we are. We now know what the universe is made of. That's a big step. We still have to explain how we know the cosmic constituents so precisely but we will get to that in chapter 4. For now we can use a pie chart to summarize what we know. Today, the slice for atoms is 5%, dark matter, 25%, and the cosmological constant, 70%. The slice for radiation is less than 0.01% and not that significant. The slices of the pie chart change as the universe evolves. Early on, radiation dominated and the other components were insignificant. Next, matter dominated. Now, in the current epoch, the cosmological constant dominates. In the future, the cosmological constant will increasingly dominate and the atoms plus dark matter will be comparatively insignificant.

We can relate these fractions to the average energy density of the universe. The whole pie corresponds to an effective mass density of about five and a half protons (or the equivalent energy computed with $E = mc^2$) per cubic meter of space. We can think of one and a half of those protons as representing all the matter (the dark matter plus the familiar matter) and the other four representing the cosmological constant. Of course there is no such thing as half a proton, but for us it just represents an amount of mass. Let's put imaginary walls around that cubic meter and let the universe expand by a factor two. The volume inside our walls increases by a factor of eight. The mass inside the walls remains the same, so its average density drops to about a fifth of a proton per cubic meter. What happens to the effective mass density that represents the cosmological constant? It stays the same at four effective protons per cubic meter! It really is like the vacuum has an energy density associated with it, which is why we count it as a component in the pie chart. We can also see why over time the cosmic pie chart becomes dominated by the cosmological constant. With a factor of two expansion over the present, the pie will have 5% for all matter and 95% for the cosmological constant.

Once we know the contents of the universe and the Hubble constant today we can determine the compactness (and temperature) of the universe throughout cosmic history. The result comes from a solution to the "Friedmann equation" which Alexander Friedmann derived from the general theory of relativity in 1922. The inputs to the equation are the matter density, radiation density, and the energy density associated with the cosmological constant; from these the equation gives the Hubble parameter versus cosmic time with today's value as a reference point. Figure 1.5 shows the solution to the Friedmann equation for a galaxy that is 110 million light-years away today. You can see that the recessional speed of the galaxy was faster 2 billion years after the Big Bang, then slowed down nearly 6 billion years due to the gravitational attraction to other matter, and is now speeding up due to the effect of the cosmological constant.

Being able to get this solution is already quite an accomplishment. However, our understanding of the universe is much more comprehensive and the cosmological model is more far-reaching than simply knowing the compactness for all times. For example, we can also understand why the universe looks the way it does, as we hope to show in the following.

2.4   Structure Formation and the Cosmic Time Line

We now combine our framework of an expanding universe from chapter 1 with our knowledge of the major components. Our goal in this section is to develop a picture for how "structure" forms. By structure we mean objects that are held together by gravity. There is a magnificent and varied array of different types of objects that range in size from stars, to galaxies, to clusters of galaxies. As can be inferred from figure 1.2 and plate 4, the space between the objects is vast and cold. In contrast, the early universe is a near uniform primordial soup of hot thermal radiation (CMB photons), electrons, protons, neutrons, neutrinos, and dark matter. How did the universe get from one state to the other? In other words, how did structure form and grow? Although we do not know the details of, say, how a galaxy forms, cosmologists have a framework that explains how there can be such a wide array of structure over such a large range of mass and how the formation process got started. Keeping in mind that this is an active area of research, we just touch on the well-established key elements of the process as it pertains to the CMB and the standard model of cosmology.

We pick up the story five minutes after the Big Bang. At this time the temperature of the universe was a little under a billion kelvin, the expansion rate was 3 million times what it is today, and the cosmological constant was irrelevant because its energy density was so much less than that of everything else. The properties of matter at this temperature, roughly 70 times hotter than the center of the Sun, are well understood. The atomic nuclear composition of the universe was roughly 75% hydrogen and 25% helium by mass. These percentages are nearly the same as they are today. This ratio was already set by nuclear reaction rates between the protons, neutrons, and neutrinos during the first three minutes, a topic to which we return later.

Initially, these primordial nuclei were in a gas with the electrons and photons, but the temperature was too hot for neutral atoms to form. Such a gas is called a "plasma," sometimes termed the fourth state of matter after solid, liquid, and gas. In the cosmological plasma, for every electron there were almost 2 billion CMB photons, and for every proton there was a little more than five times the mass in dark matter particles. Other than their participation in nuclear reactions, the neutrinos were not directly part of the formation of structure at this epoch because they barely interact with the other matter and are moving so quickly.

With the ingredients and the state of the universe in hand we now turn to the physical process. Let's put aside for a moment our concept of an expanding universe. Imagine an infinitely long one-dimensional string of equally spaced and stationary objects all of the same mass. These masses are attracted to each other gravitationally. Let's assume that gravity is the only force on them. This configuration is not stable because gravity is only attractive. Pick any mass and displace it ever so slightly to the right. Now, it is closer to its right-hand neighbor than its left-hand neighbor. Because the gravitational force is inversely proportional to the separation squared, the attraction to the right is even stronger than the initial attraction to the left: the mass and its right-hand neighbor fall toward each other. Once the spacing is changed anywhere, the whole string becomes unstable and the masses start to clump together.

The physical process behind the formation of cosmic structure is gravitational instability. We need something to get the process going, a "seed," but once it gets going, a once-uniform gas of dark matter and plasma can form structure. Of course, the one-dimensional string of masses is overly simplistic. In the full picture we have to consider all the constituents in a rapidly expanding universe. Now let's go through the process. We will put off discussing the source of the seeds until section 4.2.

There are multiple processes that take place at the same time. In one process, the seeds of structure formation initiate the clumping of the *dark* matter, but for the first 50,000 years the universe is expanding too rapidly for structure to form. In our one-dimensional analogy the masses start to fall together but the universe is expanding too quickly for them to clump. While this is happening, in a different process, the electrons and the CMB photons are strongly interacting, constantly scattering off each other. This is somewhat akin to being in a thick fog in which the light scatters off the water vapor so that every direction looks the same and you can't see very far. And, in a third process, the negatively charged electrons attract the positively charged protons (the hydrogen nuclei and in the helium nuclei) simply because opposite charges attract. The picture to have in mind is that the CMB acts most effectively with the electrons, because they are so much less massive than the protons, and the electrons pull on the protons but they can never combine to form atoms because it is too hot: the atoms would be instantly ionized. The combined interactions in the second two processes are much stronger than the force of gravity and so, *even if* the universe were not expanding so rapidly, the plasma would not clump. The electrons and thus protons are kept from clumping by the intense interaction with the radiation. Again, all three processes—clumping, scattering, and electronic attraction—are taking place at the same time.

As the universe expands, the radiation cools and the rate of expansion decreases. Soon after 50,000 years, when the matter comes to dominate the energy density, the expansion rate is slow enough for the dark matter to start clumping but it is still too hot for the plasma to clump. The interactions between the CMB and the electrons overwhelm the gravitational forces.

After 400,000 years the universe cools to the point where hydrogen *atoms* can form. In a relatively short time, the electrons bind to protons. While an electron is free, it can interact with radiation of all wavelengths, but once it is bound its interactions are restricted because it has to obey the rules of atomic physics. Shortly after the binding occurs, the electrons no longer scatter the CMB; the second and third processes mentioned above cease. Without the photon scattering, the

CHAPTER TWO

hydrogen can begin to clump. (The helium undergoes a similar process, but slightly earlier.) Agglomerations of mass already exist because clumps of dark matter have been forming since the universe was 50,000 years old. The atoms fall into the dark matter structure.

The contrast between the collections of dark matter in different regions is very small. One region may be more massive than another by a few parts in 100,000 which is like the tip of your little finger compared to your entire mass. That's all it takes to start the clumping of atoms.

The time at which hydrogen atoms form is called "decoupling" because the CMB photons decouple from, or stop interacting with, the electrons which have become bound up in atoms. The photons are then free to roam the universe. It is as though the fog has lifted and light from a distant shore can now reach you. To a reasonable approximation, the photons that land on our detectors were last scattered in this process and have since traversed the radius of the observable universe to get to us. Thus they bring to us a picture of the universe from 13.8 billion years (minus 400,000 years) ago, very much like the light from a distant galaxy brings us an image of the galaxy in its youth. The main difference is that the CMB comes to us from a time before there were stars and galaxies, from a time when matter was just beginning to form structure. This is why the CMB is sometimes called the "baby picture of the universe."

After decoupling the universe is neutral and enters a period somewhat playfully called the "dark ages" (see plate 5) because there were no stars to shine and the CMB had been cooled enough by the expansion that it did not emit visible radiation. During this time, the atoms continued their fall into concentrations of dark matter. The clumping triggered by the gravitational instability took place on all scales, from stellar sizes to huge filaments containing countless protogalaxies. The first objects to form, though, were stars. They lit up the universe, ending the dark ages.

Star formation took place about 200 million years after the Big Bang. The first generation was made of hydrogen and helium, but in their cores they produced heavier elements such as carbon, nitrogen, and oxygen through the process of nuclear fusion. These stars aged and exploded in supernovae, spewing the heavy elements throughout the universe. We are made of these heavier elements. There are ongoing searches to identify remnants and signatures of these stars; some even may have become black holes. Nevertheless, we know they have to exist because we see their ashes. More recently formed stars, such as the Sun,[7] contain elements in their surfaces heavier than helium, and these elements could not have been formed prior to the first generation of stars in the quantities we now observe. As Joni Mitchell sang in 1969, "We are stardust, billion-year-old carbon." Our knowledge of the universe has increased enormously since these lyrics were written, but "We are stardust, 13.6 billion-year-old carbon" doesn't have quite the same ring.

We also know that the first stars produced enough energy to tear the electrons from the hydrogen nuclei (the protons) by bombarding them with energetic photons. Thus, the universe began as an ionized plasma with no structure, became a neutral gas of hydrogen and helium after decoupling, and was then reionized by the first stars predominantly at an age between 500 million and one billion years. By this point though,

---

7   The Sun is 4.6 billion years old and is expected to stay in its current form for roughly another 5 billion years. It is not massive enough to produce a supernovae.

the universe had expanded enough, and the CMB was cool enough, that structure could continue to form. Nevertheless, the reionization left its mark, the newly freed electrons scattered roughly 5–8% of the CMB photons, an effect that is observed in the CMB. As with the formation of the first stars, the process of reionization is complicated and not yet well understood. It's an active area of investigation. Regardless, we know the process took place because we observe that intergalactic space is still ionized today.

As the universe ages, new stars come on the scene, galaxies begin to form, and clusters of galaxies begin to grow. The largest structures are still forming today. Although we have laid out the process sequentially, structure formation on all scales takes place to varying degrees at the same time.

The scenario of structure formation and the cosmic time line might, at first glance, seem a bit contrived. It is awfully detailed. However, the physics is straightforward and well tested. The model is predictive and there are multiple ongoing efforts to check those predictions. The picture we painted is rooted in measurements. Telescopes of all kinds are mapping out the process by looking at different structures from different epochs. If gravity acts differently than we think, if the cosmological constant isn't constant, if we don't have the correct ratio between protons and dark matter, if a new process or particle comes into play, or if neutrinos play a large role in structure formation, we can see the effects through ongoing and detailed measurements of how cosmic structure grows over time. One of the reasons for confidence in the scenario is that we know how the process started. That's one of the things we learn from the CMB anisotropy, our next topic.

COMPOSITION AND EVOLUTION

CHAPTER THREE

MAPPING THE
COSMIC MICROWAVE
BACKGROUND

WE CAN GIVE THE DETAILED ACCOUNTING WE HAVE—THE cosmic energy densities versus time, the ratio of hydrogen to helium, the epochs for different processes—because these quantities affect the CMB in characteristic and measurable ways. To understand how we can learn so much, we now focus on the small temperature differences from position to position on the night sky. The variation of temperature with position is called the *temperature anisotropy*. The word "isotropic" means "having a physical property that has the same value when measured in different directions." Anisotropic means *not* isotropic. The CMB is not isotropic, but the difference in temperature for different directions in the sky is tiny, typically one ten-thousandth of a kelvin or 0.003%.

The CMB anisotropy has been measured with exquisite precision over the entire sky by the WMAP and Planck satellites. The maps are usually shown in a Mollweide projection,

FIGURE 3.1. *Left:* A Mollweide projection map of the Earth's surface. *Credit:* Daniel R. Strebe August, 15, 2011. *Right:* A Mollweide
projection map of the CMB dipole as measured by COBE/DMR at a wavelength of 0.6 cm. Relative to our cosmic reference frame, the solar system is
traveling away from the lower-left dark region and toward the upper-right
lighter region at 0.1% the speed of light. Parts of the Milky Way's galactic
plane are just visible at this temperature range. For example, the circular
feature in the center left is the Cygnus region seen in plate 3. While both
maps use the Mollweide projection, the relative orientation on the sky of
the Earth's equator is tilted about 50° relative to the galactic plane, as can
be seen in figure 1.1. *Credit:* NASA/COBE Science Team.

which simply specifies the manner in which you represent
something that is intrinsically a spherical shell, like the Earth's
surface, on a flat piece of paper. Figure 3.1 shows the Earth in
a Mollweide projection. The equator runs horizontally along
the middle of the map, the North Pole is on top, and the
South Pole is on the bottom.

Plate 7 shows maps of the CMB anisotropy from both
WMAP and Planck. Whereas the left side of figure 3.1 is made
looking down on Earth from space, the images in plate 7 are
made looking up into the sky. These maps are oriented so that
their equators are aligned with the Milky Way.[1] The center
of the map corresponds to the center of the galaxy; the top is
the "north galactic pole" and the bottom is the "south galactic

1   The central horizontal red swaths in the panels in plate 7 correspond to the line
marked as the galactic plane, or GP, in plate 2 and to the central horizontal red swath
in plate 3.

pole." The map from WMAP was made at a wavelength of
0.5 cm (5000 microns). The one from Planck was made at a
wavelength of 0.2 cm. Both satellites made maps at multiple
wavelengths; these are just representative images. In general,
the Planck maps have higher precision than WMAPs but the
similarity in the maps once you move away from the galactic
equator is striking.

Neither WMAP nor Planck measures the absolute temperature of the CMB. If they did, the maps would be one solid
color corresponding to the temperature of the CMB. Instead,
they measure only deviations from the average temperature
of 2.725 K. The largest spatial variation is called the CMB
dipole, as shown in figure 3.1, and it too is subtracted before
producing the images in plate 7. The amplitude of the dipole
is 3350 millionths of a kelvin so, if shown, it would saturate
the color scale. The dipole arrises because the satellites have a
net velocity relative to the CMB. As you might imagine from
the Doppler effect, if you are moving toward a blackbody, it
appears slightly hotter, and if you are moving away, it appears
slightly colder. Thus, by looking around, you can tell if you
are stationary relative to the blackbody.

The existence of the dipole gives us another insight into
the universe. It means that there is a universal cosmic reference frame. This does not violate any laws of physics because
we are simply defining a reference frame. It is the same frame
relative to which galaxies are on average at rest after subtract-
ing their motion from the cosmic expansion. Most galaxies
have some peculiar velocity relative to this frame, and the
Milky Way is no exception. Our net velocity relative to the
cosmic reference frame is about 0.1% the speed of light. We
are moving quite fast with respect to the rest of the universe.
The dominant components that make up this velocity are

the motion of the Earth around the Sun (0.01% the speed of light), the motion of the Sun around the center of the Milky Way (0.08% the speed of light), the motion of the Milky Way in the Local Group, and the motion of the Local Group relative to the rest of the galaxies. Their velocities are in different directions so you have to be careful when combining them.

The component from the Earth orbiting the Sun is especially important for CMB measurements. It's called the orbital dipole and allows us to calibrate the instruments to high accuracy. Because the Earth's velocity can be measured accurately independent of the CMB, the amplitude of the orbital dipole can be predicted to similar accuracy—about 270 millionths of a kelvin. Over the course of a year, this component of the overall dipole is measured by the satellites. It is satisfying to think of calibrating variations in light from the edge of the universe with the motion of the Earth around the Sun.

In plate 7 the color bar shows the magnitude of the deviations from the average. Let's consider the regions above and below the dashed lines in the top image. Some parts are hotter than the average—the reddest places are about 2.7253 or more above absolute zero—and some are colder—the bluest places are about 2.7247 or less above absolute zero. We say "or more or less" because those colors are at the ends of the color bar. The broad hot stripe down the equator of each map is the emission from the Milky Way at these wavelengths. The map in plate 1 has had a model of this "galactic emission" subtracted. We show the full picture here so that the next time you look at the Milky Way, you can think about it at longer wavelengths and picture its relation to the CMB anisotropy. For understanding cosmology, though, the emission from the Milky Way and other galaxies is a contaminant.

THE COSMIC MICROWAVE BACKGROUND

3.1 Measuring the CMB

Before digging into the details of how we make sense of the maps in plate 7, let's go over how the measurements are actually made. The CMB was discovered by Arno Penzias and Robert Wilson in 1965. They were working at Bell Lab's Crawford Hill Laboratory, in Holmdel, New Jersey, on a telescope designed to receive signals from a communications satellite. With their state-of-the-art and well-calibrated receiver they detected an unexpected signal: that the whole sky was glowing at roughly 3.5 K. The primary reason for believing that the signal was cosmic was that it was the same in all directions.[2] Since then, the CMB has been measured with many different methods. The promise of what we might learn has driven the development of multiple new technologies and impressive instruments for measuring both its absolute temperature and the anisotropy. Here we will focus primarily on measuring the anisotropy because it's the signal from which we learn the most about the universe.

In the late 1960s, the skies were scanned with single room-temperature detectors to look for temperature differences of roughly a thousandth of a kelvin. We have now advanced to instruments with thousands of detectors cooled to just a tenth of a degree above absolute zero that run around the clock and measure CMB temperature differences at the level of a millionth of a kelvin or better. The experimental challenge is to measure these minute differences from position to position on the sky while observing with an instrument in a

---

2 Firsthand accounts of the discovery and interpretation of the CMB, including contributions from Penzias and Wilson, may be found in *Finding the Big Bang* by Peebles, Page, and Partridge, Cambridge University Press (2009). With near misses and false leads, the story shows the all-too-human path of establishing scientific fact.

300 K environment, almost a billion times hotter than the signal. The steady advances in techniques and technologies that enable us to do this have been profound and sustained.

The CMB shines over a broad range of wavelengths, although it is strongest near 0.1 cm as shown in figure 2.1. Between the wavelengths of 30 cm and 0.05 cm it is brighter than anything else in the sky if you look away from the galactic plane and observe from above the Earth's atmosphere. Especially at wavelengths shorter than 0.3 cm, water vapor in the atmosphere can make the measurements from low-elevation sites difficult if not impossible. This has driven researchers to take their instruments to high and dry locations such as White Mountain in California, the Chilean Andes, and the South Pole, or to fly them on balloons. However, the ultimate platform for measuring the CMB is a satellite.

The first satellite[3] dedicated to measuring the CMB and the infrared emission was NASA's COBE, which we discussed in section 1.1. After that came the Wilkinson Microwave Anisotropy Probe (WMAP), and most recently, Planck. We will focus on these last two, as shown in figure 3.2, as they have given us the best and most complete picture of the CMB anisotropy, and have done so in quite different ways.

Penzias and Wilson measured the CMB at a wavelength of 7.4 cm. This is in the "microwave" band and hence the name cosmic *microwave* background stuck, even though most of the emission is at shorter wavelengths. Because 7.4 cm is a relatively long wavelength it is not so affected by the atmosphere. Other common devices in the microwave band include TV stations (channels 2-83, with wavelengths 500 cm to 34 cm)

3 Physicists in the Soviet Union mounted a CMB radiometer on the Relikt satellite that came close to detecting the CMB anisotropy.

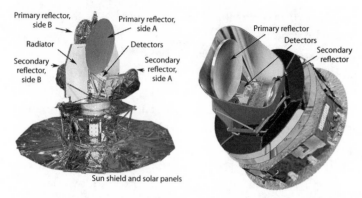

FIGURE 3.2. The WMAP satellite is on the left and the Planck satellite is on the right. For scale, the large reflector or "dish" on WMAP is 140 cm by 160 cm. For Planck it is 150 cm by 190 cm. The satellites are comparable in size with overall heights about 300 cm. For WMAP the sun is in the direction of the bottom of the page, and for Planck the Sun is off to the lower right. The thermal shielding allows WMAP's primary reflector to cool to 60 K and Planck's to below 40 K. For both satellites, the detecting elements are just below the primary reflectors. For WMAP, the detectors are passively cooled to 90 K by removing the heat from the detectors with the large radiators. Planck uses an active system to cool the longer wavelength set of detectors to 20 K and the shorter wavelength ones to 0.1 K. *Credit:* ESA and the Planck Collaboration; NASA/WMAP Science Team.

and microwave ovens (12.2 cm). Modern TV satellite dishes operate near a wavelength of 1 cm. You can see in figure 3.2 that WMAP looks like it has two back-to-back TV satellite dishes. This is no coincidence. It operates between the wavelengths of 1.3 cm and 0.3 cm in five different bands.

To get a better sense for how the measurement is done, we can think about an old-fashioned TV. Let's say you have the type of antenna that attaches directly to your TV. If you then tune to channel 83 and there is no broadcast there, you will just get fuzz or noise on your TV screen. This fuzz comes from two sources. It is a combination of microwaves coming into the TV from the environment through the antenna plus noise from the electronics inside the TV. Consider the antenna component of the fuzz. The incident microwaves move the electrons in the antenna structure. In turn, those electrons tickle the input of the transistors in the TV receiver and the rest of the TV receiver amplifies and packages the signal so you can visualize it. The CMB enters the antenna just like a broadcast signal would, but it looks like noise. To a crude approximation, about 1% of the total noise on the TV screen is from the CMB that enters through the antenna.

To measure the anisotropy you would point the TV antenna in a specific direction and record the amount of fuzz, say, by taking a picture of it or recording its hiss. Without changing anything on the TV, you'd then point the antenna to a new location and again record the amount of fuzz. The difference in the amount of fuzz directly corresponds to the difference in temperature of the radiation coming into the antenna.

It is easy to imagine how to improve the measurement. You'd definitely want to get a less noisy TV receiver so that more of the noise came from the CMB. It is somewhat counterintuitive but it is not necessary to make your instrument colder than the CMB to measure it. The key is that the electrons in the detecting elements are free to respond to the CMB. If the transistors are cooled to, say, 100 K, the electrons are freer and the electrical noise from the transistors is reduced. You could increase the signal by listening in multiple channels at once. You'd want to make sure that your antenna only accepted TV waves from directions close to where you were pointing it. You'd want to make a TV that worked at shorter wavelengths, etc. In effect WMAP does all of this, although with very fancy transistors and a lot of attention to detail.

WMAP's other key feature is that it accepts radiation from two different directions simultaneously and compares them. This is why there are back-to-back dishes. The instrument is not capable of measuring the absolute temperature, only a large collection of temperature differences. Then, a computer program takes all those difference measurements and combines them to produce a map of just the spatial variations as shown in plate 7.

The Planck satellite, shown in figure 3.2, takes a different approach. It has just one primary receiving dish. The idea is to spin the satellite, find the average temperature over the spin, and look at temperature differences around that. In total there are 72 independent detection channels measuring the sky at any time, as opposed to WMAP's 20, but in both there is quite a bit of redundancy. Planck has two instruments that in combination measure between the wavelengths of 1 cm and 0.035 cm in nine different bands. The lower three bands are similar to WMAP's, but the upper six are different and use a different technology called bolometry.

The bolometer, a fantastically sensitive device, is a lot like a thermometer. It simply measures the amount of thermal energy dumped onto it. Unlike transistors, bolometers do need to be fairly cold to detect the CMB. The key to using them is to isolate them so well that only the radiation you intend gets to them. The ones on Planck are cooled to 0.1 K above absolute zero. In one second they can measure a temperature difference smaller than one ten thousandth of a kelvin.

Both satellites observed from "the second Lagrange point" or L2. This is a location in the solar system that is about a million miles away from the Earth in the direction opposite the Sun. In 1772, Joseph-Louis Lagrange found there were five places in the solar system where the gravitational pull of the Earth and Sun balanced in just the right way to make an orbit possible. The L2 position is unstable, so miniature jets on the satellite are fired ocasionally to keep it from wandering too far away. Unlike most satellites, WMAP and Planck orbit the Sun and not the Earth.

From L2, the dishes of both satellites generally look away from the Sun, Earth, and Moon. This is important because these bodies are hot compared to the tiny temperature we need to measure. Another important feature of L2 is that it is thermally stable. There are no day/night cycles. This stability is important for being able to measure the sky over and over and then averaging the data together. WMAP observed for nine years; Planck for four years.

Although the function of the satellites is straightforward—they simply measure the radiation temperature of the sky—getting them to work to the limits they have achieved has required unprecedented control of systematic sources of error. The devil is in the details. For example, you have to be sure that a measurement made on one day can be directly compared to a measurement made two years later at a level set by the fundamental noise characteristics of the instrument. By far the most computationally intensive part of the data analysis is checking that you understand the instrument and the environmental impacts on it.

One of the recurring questions when measuring the CMB is: "how do you know that you're really looking to the farthest reaches of light and not just at something in the Milky Way or in the Local Group?" The primary way to determine this is by measuring the anisotropy at different wavelengths. Just as Planck's equation precisely describes the amount of power per wavelength band for a blackbody, it prescribes a similar relation for the fluctuations. Emission from the galaxy has a much

68

different pattern of emitted power versus wavelength so it can be distinguished from the CMB. Both Planck and WMAP have multiple wavelength bands that make it possible to clearly separate the "foreground emission" from the cosmic emission.

A straightforward way of separating galactic emission from the CMB, sufficient for our purposes, is to simply "mask out" the galaxy. This means that you remove, or mask, that part of the map from any analysis. In plate 7 you can do that in your mind's eye by excluding regions between latitudes plus and minus 20° as indicated by the dashed lines on the top map. Northward and southward of this region, the maps show a seemingly random array of hot and cold regions of various irregular shapes and sizes. This is the CMB anisotropy signal we are after.

When looking at the Hubble Ultra Deep Field in plate 4, you might wonder why CMB experiments don't see all of those galaxies. There are three reasons for this: those galaxies emit radiation at different wavelengths, there is a lot of space between the galaxies (most of the image is black), and the galaxies are small in angular extent. When we measure the sky with CMB telescopes with high angular resolution, we can see galaxies and clusters of galaxies and thereby determine that their contribution to the maps in plate 7 is practically negligible.

We show two different maps in plate 7 to make an important point. The anisotropy is measured with high precision and is the same as determined with two completely independent satellites with different detectors, observing strategies, and scientists. The data were processed by independent and competing groups, yet, ultimately, the two maps show the same thing. The results are confirmed.

Measuring the CMB is now "big science." Up until the 1990s, groups of two or three researchers could make a breakthrough measurement with inexpensive equipment. Now there are thousands of people in the field and the instruments cost many millions of dollars, or hundreds of millions for a satellite. The satellite maps will remain the most accurate full-sky measurements, but in select regions of sky and at certain angular scales you can improve on them significantly. A network of ground based telescopes is already in the process of mapping the CMB over half the sky with even greater precision than Planck or WMAP.

3.2 The CMB Anisotropy

Let's say we have a map from either WMAP or Planck that has been masked or cleaned such that we're confident the remaining signal is the CMB anisotropy. It's just a collection of temperatures, a heat map. How do we learn cosmology from it? First, recall that these maps provide a picture of the edge of the observable universe from 400,000 years after the Big Bang. Although the universe went through decoupling throughout its entire volume at this time, we can think of this radiation as coming to us from a surrounding surface because that is the region from which the CMB we now measure originated. The region is sometimes called the "decoupling surface" to remind us that the light we detect decoupled from the primordial plasma there. In plate 5 the decoupling surface is the outermost shell.

What do the hot and cold regions show us? The connection we want to develop is that they trace out a map of the strength of gravity in the universe just 400,000 years after the Big Bang. It will take a few steps, but the connection is important because it will allow us to relate the spatial distribution of matter to the temperature anisotropy of the CMB.

THE COSMIC MICROWAVE BACKGROUND

Earlier we gave a simple one-dimensional picture for how mass clumps to form structure. In the universe, that of course happens in three dimensions. When the mass clumps, the strength of gravity in that region of space is stronger than in others. If, say, the Earth were the same size but more massive, we would weigh more because the force of gravity would be stronger. Similarly, the more mass that clumps into a fixed volume, the greater the strength of gravity. We call this variation in the strength of gravity throughout space the "gravitational landscape." In turn, the gravitational landscape produces the CMB anisotropy, as we describe next.

It is sometimes easier to think of the clumping on a two-dimensional slice through space. We can imagine a two-dimensional section of land with hills and valleys of all sorts of different widths and heights. The different heights of the land represent the different strengths of gravity. Gravity is stronger in the valleys than on the hilltops. As the universe evolves prior to decoupling, the dark matter clumps, and the valleys get deeper. The plasma of photons, electrons, and nuclei tries to fall into the valleys, but it is so energetic that it doesn't clump.

There aren't really terrestrial analogues for the process. Loosely one can think of the plasma like rather agitated water trying to settle into an array of plastic egg crates. The crates represent the hills and valleys. Unlike water, though, the plasma is compressible. When it falls into a valley it compresses, heats up, and then bounces back.

The whole universe at this time is full of a compressing, rarefying, bouncing, oscillating plasma that is trying to collect in the valleys but can't settle down. Then, in a relatively short time, the universe cools enough for atoms to form, the CMB is set free, and the plasma state ends. This is the decoupling at 400,000 years. The CMB records the state of the universe at this time. The plasma that has fallen into a valley is compressed and heats up, somewhat like gas compressed in a piston. To a good approximation, the hot regions, the red in the maps, show the locations of the gravitational valleys where the plasma is hotter, and the blue regions show us the locations of the hills. The CMB provides a snapshot of the primordial gravitational landscape. The freed up atoms then go on and respond to this landscape to gravitationally collapse further into the cosmic structure discussed earlier.

Typically the level of contrast in the maps, that is, the difference in temperature in different regions, is 100 millionths of a kelvin or 100 microkelvin. We often use the term "fluctuations" as shorthand for "variations in space" and say that the temperature fluctuations are roughly a few parts in 100,000 of the total, 3 K. This is the same fraction as for the clumping in the matter that we discussed in the last chapter. If your mass represents the amount of clumping at decoupling, one region might have a copy of you and another might have a copy of you plus or minus the mass in the tip of your little finger. The clumping of matter and the level of the anisotropy are intimately related—indeed, the source of the fluctuations is the same: the primordial seeds we mentioned earlier.

Plate 8a shows a close-up picture of the region in the small gray box in the top map of plate 7. It is about eight full-moon diameters on a side. Although the hot and cold regions are splotchy and irregularly shaped, they do have a characteristic size. The image clearly does not look like a pointillist painting made up of thousands of tiny patches of color. Nor are the color patches so large that they take up half the image. The characteristic patch size looks like roughly twice the diameter of the full moon.

THE COSMIC MICROWAVE BACKGROUND

Why is there a characteristic splotch size? Let's go back to the gravitational landscape. The primordial plasma will be hottest when it is compressed the most. This happens when the plasma "flows" down the valley walls from all sides just once in the time between when it starts flowing, roughly 50,000 years after the Big Bang, and decoupling. The flow speed is more properly thought of as the speed of a disturbance in the plasma, similar to how sound travels in air. The flow speed of the plasma is fixed by fundamental physics. The time over which it can flow is set by the expansion of the universe because after 400,000 years there is no longer a plasma and the decoupled CMB is free to travel unimpeded. The flow *speed* multiplied by a *time* is a distance. Therefore, a special size of valley exists that is particularly effective at creating hot splotches. Similarly, the same special size of hill makes the cold splotches. The speed of the plasma and the optimal valley size may be computed theoretically to high accuracy using conventional physics. To be sure there are valleys of all sizes and depths, but the CMB highlights a special size and this size may be computed in units of light-years.

We can make a simple approximation of the special size. The time over which the plasma can flow is $400,000 - 50,000 = 350,000$ years. The speed of the plasma is more difficult to compute. Because the plasma is predominantly composed of photons, the speed of a disturbance in the plasma is very fast, about half the speed of light. Multiplying these numbers together we get around 200,000 light-years, which corresponds to the distance between the bottom of the valley and the edge. This isn't quite correct because during this time the universe expands by a factor of three (see appendix A.3). A more detailed calculation reveals that the size is closer to 450,000 light-years. This is called the characteristic acoustic scale at decoupling because the plasma flow is like a sound wave. Our special size, corresponding to the *diameter* of, say, a hot or cold spot, is twice this, or about nine times the diameter of the Milky Way as measured today.

We can now understand what the universe was like back then relative to today. It was a thousand times hotter and much more uniform. If we divided space into volumes of 900,000 light-years on a side, some would have more mass than the average by a few parts in 100,000 and others would have less by the same fraction. This tiny mass difference, traced out by the CMB anisotropy, grows through gravitational instability to form cosmic structure while the universe is expanding.

This process was first spelled out in the 1970s by Jim Peebles and Jer Yu and in a related paper by Rashid Sunyaev and Yakob Zel'dovich . The model has been refined and augmented over the decades, but the basic picture we have today is the same. It is a testament to the universality of physics that predictions can be made for what should happen in the early universe based on measurements made on Earth, and that those predictions can be tested. The physics behind the hot and cold regions is more involved than we described above, but the process we presented is the dominant one that gives rise to the features in the CMB maps.

3.3 Quantifying the CMB

To compare with theoretical models, we need to quantify the maps. In other words, we need to reduce the randomly spaced set of hot and cold regions of different temperatures and irregular sizes to a set of numbers. Mathematically, the anisotropy maps are two-dimensional sets of random numbers on a sphere. Over the decades a few methods have been developed to characterize maps like these. We will look at two.

The first method is simple and powerful. We simply go through the map and everywhere there is a hot spot we extract a 4° × 4° section of map centered on the hot spot. We have to be careful that we do not double count but we can experiment and devise an algorithm for that. On the left of plate 8a, we see that there are about a dozen hot spots, so in the area around this region we'd make a dozen 4° × 4° patches. Over the region of the sky well away from the galactic plane, north and south of the dashed lines in the top of plate 7, there are about 10,000 hot spots. We then take all those 4° × 4° maps and average them together. Through this process, the average hot spot emerges while the features in the maps that are not common average away. Although we focused on the hot regions, the same could be done to the cold regions.

The right side of plate 8a shows the average hot spot map for Planck. This is an amazing picture. It is showing us that special valley size in enormous detail. To the eye it looks to be roughly two full moons across. A more detailed analysis gives the angular diameter as 1.193°, which we will round to 1.2°. Along with the CMB temperature, this is one of the most precisely measured numbers in cosmology. It has far-reaching consequences, to which we will return later.

The second method for making sense of the maps is more involved but it displays the details of the spot more clearly. This results in a plot called a "power spectrum" as shown in figure 3.3. In essence the plot tells us the magnitude of the fluctuations of the temperature in the map for different angular sizes. We already know from the above that the largest fluctuations in temperature will be for regions around a degree in size, which corresponds to the maximum in figure 3.3 near 1°.

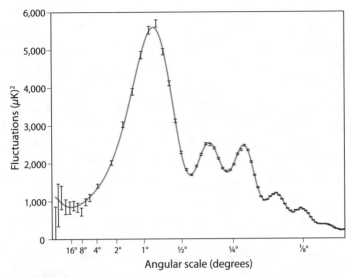

FIGURE 3.3. The power spectrum of the CMB anisotropy from WMAP and Planck. The variance or magnitude of the fluctuations is on the y-axis, and the angular size is on the x-axis. The angular scale markers are not evenly spaced and decrease by half with each tick. The maximum is near 1° roughly corresponding to the diameter of the hot spot on the right side of plate 8a. The gray line shows the best fit model based on the six parameters that describe our universe. The measurement uncertainties are indicated by the vertical black lines at each point.

One way to think about this plot is as a graphic equalizer, or simply "equalizer," for a fancy audio system. This is a device that lets you amplify or soften different frequencies of sound. For instance, you might want to emphasize the bass notes over the treble notes in a piece of music. A car radio might do this with one "tone" knob, but an equalizer lets you do it with fine control. A typical equalizer has five to ten columns of little lights. The number of lights lit up in the left-most column indicates the loudness of the bass notes; the lit-up ones on the right tell you about the treble. In our analogy, the map of the CMB corresponds to the music. The bass notes are on the left side of the plot and the treble notes on the right, arranged similar to a piano keyboard. The y-axis then corresponds to the loudness of each audio frequency or pitch. If the peak near 1° corresponds to middle C at 261 Hz, the second peak would be at 635 Hz or just below E in the next octave higher, and the third peak would be at 963 Hz in this same octave but just below B. In the map, these second and third peaks are not easily discernible to the eye, but they are there. Somewhat prosaically, the plot thus shows us the music of the cosmos. Or, more accurately, it shows us the harmonic content of the cosmos.

Figure 3.3 is one of the most important plots in cosmology. It is the culmination of more than five decades of work by scientists from around the world. At the start of the quest to make the plot, no one knew what we would find or even how much we could learn once the measurement was made. We now interpret every little bump and wiggle in detail. Later we'll interpret the full plot, but to give you an appreciation of what it has to offer, we can determine the composition of the universe from the positions and amplitudes of the peaks.

The plot is so important that we'll consider it from another angle. While the equalizer analogy interprets the plot in musical terms, the plot is really telling us about fluctuations in a two-dimensional map of the CMB. Let's think about the spatial aspect. Imagine that you are far from the beach looking out on the ocean. Think of what you would see if you immediately froze the pattern on the surface. In this frozen seascape you'd see large swells, medium-size waves, and small ripples on top of them. The frigid expanse is like the map of the anisotropy with the heights of the frozen water representing the temperature fluctuations in the CMB. The average depth of the ocean could represent the average CMB temperature of 2.725 K. In that frozen ocean, the swells have the longest wavelengths and the largest heights (if you are way out in the ocean and not in a storm), the waves have shorter wavelengths and medium-size heights, and the ripples have the shortest wavelengths and the smallest heights. Now let's say we got in a plane and looked at our frozen ocean from above. The angular separation of the peaks of the swells would be larger than the angular separation of the peaks of the waves and would be larger still than the angular separation of the peaks of the ripples. On a power spectrum plot, the swells would be on the left side of the x-axis at large angles and in our example they'd have a high value on the y-axis. The waves would be in the middle of the x-axis with a medium value on the y-axis, and the ripples would be on the right with the smallest value on the y-axis. You can see the effectiveness of this type of plot. In a compact way it tells you the characteristics of all the different fluctuations—swells, waves, ripples, etc.— on a frozen ocean. Conceptually, finding the power spectrum of the CMB is not too different from finding it for the ocean surface. The peak in the CMB plot near 1° would correspond to unusually high frozen waves of this angular size as viewed from our plane. We shouldn't push the analogy too far, though, because the CMB fluctuations are random, whereas the ocean's fluctuations, that is the waves, are not.

Last, let's think about the plot in terms of how we might make it. The actual procedure involves specialized algorithms and, like the measurements, has evolved and matured over the years. However, it is not too difficult to get an operational sense for how the algorithms work. The following details are not important for other sections but ideally will give more insight into figure 3.3. To start, take a map, cut out the region contaminated by the Milky Way, and then cut the remainder of the map into disks, say, 8° across (16 full moons). For each of those 8° diameter disks, compute the *average* temperature. Of course, in an 8° disk there will be lots of smaller hot and cold regions, but they will average out. You'll end up with a set of *average* temperatures for all the 8° disks. Some will be hotter than zero, others colder. We don't so much care about the average disk temperatures; we just want to know how much they scatter around zero. The common way to determine this is to subtract the average of all the disk temperatures from each individual disk, square the remaining part of each disk's temperature, because this makes them all positive, and then average those. This is called a "variance" and is why the y-axis of the plot has units of $(\mu K)^2$. Now, repeat the process for a list of, say, 100 disk sizes ranging from 16° in diameter all the way down to 1/8° in diameter. Then, because a smaller disk will have all the variance of the next larger disk and then some, you have to go through and subtract entry 99 from entry 100, subtract entry 98 from entry 99, and so on. You'll end up with a new list that has the variance associated just with each disk's angular size. Finally, go through the list and multiply each number by the disk's angular size. You then plot the entry in the new list on the y-axis and the disk diameter on the x-axis. In broad outline, the resulting figure will resemble figure 3.3 but without all the details.

In figure 3.3 we see there is more going on than just the degree-scale fluctuations. The other ups and downs come from different ways the plasma oscillates and interacts with the gravitational landscape. For each data point in figure 3.3 there is an uncertainty represented by the vertical "error bar." The smooth line that goes through the data is the standard model

PLATE 1. This is a map of the variations in temperature of the remnant light from the birth of the universe as measured across the full sky. The temperature difference between the bluest and reddest regions corresponds to 400 millionths of a degree Celsius. The goal of this book is to explain this image and what it tells us about the universe. (*Credit*: NASA/WMAP Science Team)

PLATE 2. The Milky Way, running diagonally upward at roughly 45° from the lower left. This picture, by Giulio Ercolani and Alessandro Schillaci, was taken from just outside of San Pedro de Atacama, Chile. Cerro Licancabur is in the lower left. The bright dots away from the galactic plane are stars in our galaxy. The galactic plane lies between the arrows on the periphery labeled "GP." The intersection of the galactic plane and the line indicated by "GC" marks the galactic center. The dark regions in the galactic plane are "dust lanes." The dust blocks visible light but emits thermal radiation, as shown in plate 3.

PLATE 3. The glowing dust in the Milky Way as observed by the DIRBE instrument aboard the COBE satellite. This image shows far-infrared radiation at a wavelength of 100 microns or about 200 times longer than that for plate 2. While in plate 2 the dust obscures starlight, at this wavelength you see the dust glowing. The center of the Milky Way is in the center of the image. The blob above the center is the Ophiuchus Complex, a large dust cloud. The feature on the far left is the Cygnus region, and the bright spot to the lower right is the Large Magellanic Cloud, a nearby dwarf galaxy. (In plate 2, the Large Magellanic Cloud was off to the right and below the horizon when that picture was taken.) This picture covers a quarter of the sky. If you were out in the desert and could see the Milky Way at these wavelengths, the image width would span from horizon to horizon. (*Credit*: NASA/COBE Science Team)

PLATE 4. The Hubble Ultra Deep Field. The vast majority of the objects in this image are galaxies. The light from the nearer ones has been traveling to us for a billion years; the light from the farthest ones for about 13 billion years. This picture was taken while observing in the direction of the constellation Fornax. (*Credit*: NASA, ESA, and S. Beckwith (STScI) and the HUDF Team)

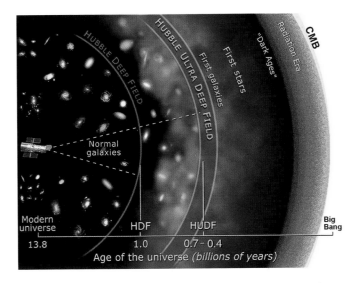

Normal galaxies

Modern universe

13.8   HDF   HUDF   Big Bang
         1.0   0.7 - 0.4

HUBBLE DEEP FIELD

HUBBLE ULTRA DEEP FIELD

First galaxies

First stars

"Dark Ages"

Radiation Era

CMB

Age of the universe *(billions of years)*

PLATE 5. Telescopes are like time machines. As we look out in space we look back in time. With the Hubble Ultra Deep Field image we look back to when the galaxies began to form. Light from the first stars was emitted when the universe was roughly 200 million years old and has been traveling to us since then. We can think of it as coming from a shell out near the edge of the observable universe. The CMB comes to us from a shell essentially at the edge of the observable universe. In this picture the CMB is the outer yellow ring. The label "Big Bang" marks the beginning of the time line. (*Credit*: NASA)

PLATE 6. A composite image of the Bullet Cluster made with data from the Chandra X-ray Telescope, the Magellan Telescope, and the Hubble Space Telescope. The image is about one-sixth of the full Moon across. The white and yellowish objects are mostly galaxies. The pink areas show the normal matter, which is primarily in the form of a hot X-ray emitting gas. The blue areas show the location of the dark matter as revealed by gravitational lensing. Note the concentration of galaxies in the blue regions. (*Credit*: X-ray—NASA/CXC/CfA/M. Markevitch et al.; Optical—NASA/STScI; Magellan/U. Arizona/D. Clowe et al.; Lensing Map—NASA/STScI; ESO WFI; Magellan/U. Arizona/D. Clowe et al.)

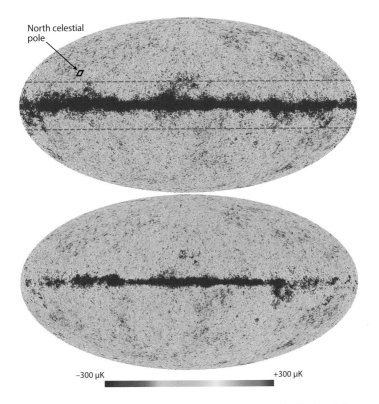

North celestial
pole

−300 μK　　　　　　　　　　+300 μK

PLATE 7. Maps of the full sky showing the CMB anisotropy and galactic emission in a Mollweide projection. *Top*: The Planck map at a wavelength of 0.2 cm. Radiation from the relatively nearby Milky Way is primarily between the dashed lines. Most of the signal above and below these lines is the CMB anisotropy, although in a few places the Milky Way emission pokes through. The little square box on the left just above the top dashed line is centered on the North Star and shown in plate 8a. *Bottom*: WMAP map. The features in the two maps are the same once you get away from the Milky Way. The temperature color scale runs from −300 millionths of a degree to +300 millionths of a degree. The "μ" sign means "millionth." (*Credit*: ESA and the Planck Collaboration; NASA/WMAP Science Team)

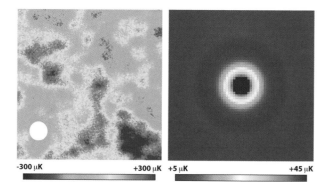

-300 μK   +300 μK   +5 μK   +45 μK

PLATE 8A. *Left*: A close-up of a 4° × 4° section of the Planck map in plate 7 centered on the north celestial pole. For scale, the white circle shows the size of the full moon. Of course, the full moon is not near the North Star. *Right*: The average of more than 10,000 hot (or red) spots in the Planck map also shown as a 4° × 4° image. The irregularities of individual spots average out. On this plot, blue is near the average of all hot and cold spots, 2.725 K, whereas red, the average hot spot temperature, is 45 μK greater than that. (*Credit*: ESA and the Planck Collaboration)

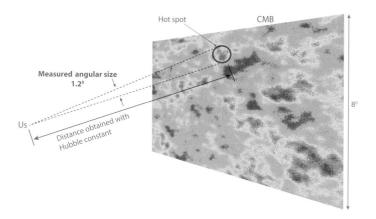

PLATE 8B. A depiction of measuring the size of the hot and cold spots in the CMB. This average hot spot size is shown in plate 8a and corresponds to the peak of the power spectrum in figure 3.3. By combining the measured angle and the computed size of the spots with knowledge of the Hubble constant, we can determine the geometry of the universe. (*Credit*: ESA and the Planck Collaboration)

of cosmology. You can now see why measurements of the CMB anisotropy are so powerful. They are extremely accurate and highly constraining. Any potential theoretical model of the universe has to fit these data. If the model doesn't fit, it is ruled out. If a model cannot make a prediction for this plot, it is not a contender. You can also see why cosmologists are confident that we have the basic picture correct even though we don't know all the elements of the model in depth. In the next section, we discuss how the model relates to the gray curve.

Before moving on, let's do a quick thought experiment to put our map of the CMB anisotropy in a broader perspective. Let's imagine that we could live 13.8 billion years and witness all of cosmic history. You might wonder, where did the Big Bang happen? It happened everywhere at the same time. In particular, it happened right were we are. Of course the universe was much more dense then but still, as far as we are concerned, infinite. If we started our stopwatches just after the Big Bang, we would experience the formation of the light element nuclei at 3 minutes, decoupling at 400,000 years, the formation of first stars at 200 million years, and so forth. After decoupling took place, the CMB photons, free at last, had 13.8 billion years to travel to the edge of the observable universe. Around us here on Earth, there was a valley in the gravitational landscape, so that the Local Group (figure 1.2), including the Milky Way, could form. The same physical processes took place everywhere in space at the same time, though some locations were in the bottoms of valleys, others at the tops of hills, and most others somewhere in between.

Now imagine that you were instantaneously transported today to the edge of the observable universe and looked back toward Earth. What would you see? The galactic environment around you would be similar to the environment we see around us now. Recall that at any fixed age, the universe looks the same everywhere. Your environment would be different in terms of the particulars—that is, you would see galaxies that we cannot see from Earth—but it would look the same on average. As you looked back towards Earth and our Local Group you might see a CMB hot region because, as we know, matter had to clump in our vicinity to make the Local Group. We say "might" because the Local Group would appear small in angular extent compared to a typical CMB fluctuation. But you would not see any galaxies in the Local Group because the light from them would not have reached you yet.

Now that we have a sense of the bigger picture, and how to think about the observational measurements of the cosmos in a physically intuitive way, it is time to change gears and introduce the major theoretical elements of the standard cosmological model. This will require more advanced concepts from physics, and you may have to take a bit more on faith. However, the reward is that we will reach a description of the six cosmological parameters that characterize our universe and all the measurements made so far of its large-scale properties. We begin the next chapter by considering the geometry of the universe.

CHAPTER THREE

CHAPTER FOUR

THE STANDARD MODEL OF COSMOLOGY

4.1 The Geometry of the Universe

ONE OF THE FUNDAMENTAL CHARACTERISTICS OF THE universe is its geometry. Geometry is the study of the relations between points, lines, angles, surfaces, etc. Let's go back to thinking about space. We have noted that it can expand at different rates. It is also malleable and can be warped or curved, as we saw with gravitational lensing by the Bullet Cluster. For lensing, it was not too much of a stretch to think of the space near a massive object as being curved. But now we want to think about our whole three-dimensional space as being curved into a fourth spatial dimension. This is a little more challenging.

In the mid-1800s, Georg Friedrich Bernhard Reimann showed that even without going to the next higher dimension we can tell if we live in a curved space. To understand his insight we will work with two-dimensional surfaces

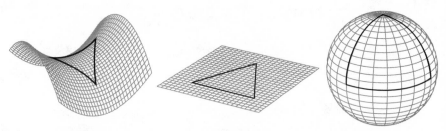

FIGURE 4.1. Examples of possible geometries of two-dimensional space. On the left is the saddle-surface-like open geometry. Think of it going on forever. The thick dark lines show a triangle whose interior angles sum to less than 180°. The middle shows a flat sheet-of-paper-like geometry. Think of it too as going on forever. Here the thick dark lines show a triangle whose interior angles sum up to 180°. On the right is a spherical-surface-like geometry. This one is finite. Here the thick dark lines show a triangle whose interior angles sum up to greater than 180°.

curved into our familiar three-dimensional space, as shown in figure 4.1. Imagine that you are an ant walking around on the two-dimensional surface between the three vertices of a triangle. Think of the surface as being very large, the ant as very small with negligible height, and assume that all motion is confined to the surface. If you walk the perimeter of *any* triangle on a flat piece of paper and sum up the interior angles, you will get 180°. The piece of paper is said to have a "flat" geometry and the two-dimensional space is infinite so that there are no edges.[1] By convention, we use the word "flat" even if we make the triangles in three-dimensional space with all possible orientations as opposed to on a flat sheet of paper.

1 A Möbius strip is an example of a "flat" geometry that is not infinite. It is said to have a non-trivial *topology*. Topology describes the way in which space is connected. For example, a doughnut has a different topology than a sphere because you can't deform one into the other. In this book we assume that the universe is characterized by its geometry and not its topology. Predictions from cosmologies with non-trivial topologies can be tested with CMB maps. So far, there is no strong evidence for a non-trivial topology.

Let's consider instead a spherical shell. This is an example of a finite and closed, positively curved space. The ant could walk a triangular path from the North Pole to the equator, around the equator by a quarter of the circumference, and then back to the North Pole. Now the ant would find the sum of the interior angles to be greater than 180°, and for this particular path the sum would be 270°. The larger the triangle, the larger the sum of interior angles. If the ant shot a laser beam out in this space, it would come back and hit his back side because in two dimensions the laser beam is confined to follow the surface of the sphere.

A horse saddle is an example of an open, negatively curved space. Unlike the case for the spherical shell, the saddle surface, like the flat piece of paper, goes on forever. If the ant were to walk the perimeter of a triangle on the saddle surface and sum up the interior angles, he would find that they were less than 180°. If we think of the leather as space and if we tried to flatten the saddle onto a flat surface, we would have a bunch of folds left over. An open, negatively curved space is one in which there is more available space, more leather, the farther away we go.

This same process of determining the overall geometry by measuring the sum of the interior angles of a triangle on a two-dimensional surface also works for a three-dimensional space that is curved into a *fourth* spatial dimension. All we have to do to figure out the geometry of our three-dimensional space is to traverse a large triangular path and sum up the angles. Locally we might do this by making a triangle out of the Sun, Moon, and Earth and measuring the interior angles, but the best test is to make a very large triangle. The CMB gives us a way to do this, but on a cosmic scale.

THE STANDARD MODEL OF COSMOLOGY

You may recall from high school geometry that to determine all the angles in a triangle, you need three pieces of information. They could be, say, the lengths of two sides and one of the angles. The CMB hot or cold spots form one side of the triangle. In effect we average all the spot sizes together. As we discussed in the previous section, we can compute to high accuracy the average physical size of a hot or cold spot in, say, light-years. Using our maps we can measure the average *angular* size to high accuracy as shown in figure 3.3 and plates 8a and 8b. We need to know one more bit of information to completely specify the triangle that has, say, a hot spot as the far side. Although it is not obvious, that piece of information is the Hubble constant because it links the physical size of the hot spot to the distance to the hot spot. The result of the accounting is that the sum of the interior angles is 180°. To the limit of measurement, the geometry of the universe is "flat."

A simple calculation can help us get a sense for how the pieces fit together. Earlier we found that the *computed* size of a hot spot is about 900,000 light-years across at decoupling. The universe has expanded almost a factor of 1100 since then, so today this is 990 million light-years across. Its angular size is *measured* to be about 1.2° across by the Planck and WMAP satellites. From this we compute the distance to the decoupling surface to be about 46 billion light-years, which we recognize as the radius of the observable universe.[2] If the geometry of the universe had been closed, we would have measured a larger

---

2   For a flat geometry, the ratio of an object's angular size to the 360° in a circle is the same as the ratio of the object's physical size in, say, light-years to the circumference of the circle also in light-years. For a hot spot, this translates to 1.2°/360° being the same as (hot spot size)/(2π distance). After plugging in numbers we find a distance of 47.2 billion light-years. Had we used less approximate numbers, the agreement with 46 billion light-years would be even better.

angular size for the spots; if it had been open, we would have measured them to be smaller. It all hangs together!

To summarize, the geometry of the universe is like the geometry many of us learned in high school. It is the simplest one we can think of. It is what you would have expected if you had never heard of Einstein or Reimann. More important, the geometry has been determined by measurement and can be checked with different types of measurements of the universe.

4.2 The Seeds of Structure Formation

The very very earliest instants of the universe are still not well understood. The reason is that as yet there is no fundamental theory that combines gravity with the standard model of particle physics. In place of this we have "effective theories" and paradigms that are deeply rooted in the physics we know and that can explain the observations. The best known of these is "inflation." We touch on aspects of it now, but keep in mind that this is still a very active area of theoretical research.

One of the great mysteries of the cosmos before the inflation model was invented was: "why are the properties of the universe in two opposite directions so similar?" To be concrete, let's take the north celestial pole and south celestial pole as our opposite directions and consider how the CMB can have the same temperature in the two directions. According to the picture we have been developing, the light from both sides of the observable universe is just reaching us now. No information can travel faster than light, so there is no way that radiation from the north celestial polar direction could have passed us and gone on to affect what we see in the south celestial polar direction, and vice versa. Yet, they are nearly the same temperature, 2.725 K, and have the same properties.

In the inflation model, the space in the very early universe, before there were *any* particles, had an enormous energy density. Associated with this energy density was a pressure that made space, or caused an expansion, at an unimaginably fast, exponential pace. At the start of the process imagine you have two regions, call them Alice and Bob, that are right next to each other and that share information. In inflation, space between Alice and Bob is made so rapidly that they can no longer communicate with each other. Their *apparent* speed of separation is faster than the speed of light. They are separated beyond, perhaps many many times beyond, the distance over which they can subsequently affect each other.

Inflation takes place over an extremely short period of time, very roughly a billionth of a billionth of a billionth of a billionth of a second. After inflation ends, the universe settles into a calmer pace of expansion. As the universe ages, the observable universe gets larger and larger because we can look farther and farther away. At some point, Alice and Bob come into view with Alice, say, in the north polar direction and Bob in the south. Now we have a mechanism for saying why opposite sides of the universe might look the same. They communicated with each other very very early on, became hugely separated during the inflation epoch, and are just now coming into our observable universe. We also need a mechanism to explain why they would not separately evolve when they are out of our sight, but that too is part of the model.

There are many variants of inflation, but the simplest model has two other features that are relevant to the CMB. The first is that the universe is geometrically flat to at least a part in 10,000. This corresponds to trying to determine if a meter stick is flat or bowed up at one end by 100 microns. As the inflation model was proposed prior to the observations, it gained a lot of credibility when the data showed the universe had a flat geometry. The idea is that even if the earliest geometry were, say, positively curved, inflation would have expanded it so much that it would effectively be flat. It is not difficult to imagine this in two dimensions. If you are on the surface of a sphere, such as the Earth, you can tell the surface is curved. But, if the radius were a billion billion times larger, it would be difficult to tell you were on a sphere. To the limits of measurement, our geometry is flat but we cannot rule out the possibility that it is just ever so slightly positively or negatively curved.

The second feature is that inflation incorporates a mechanism for generating the seeds for the formation of cosmic structure. The seeds are quantum fluctuations in the primordial energy density. What do we mean by quantum fluctuations? Think of them as tiny localized fluctuations in energy on the subatomic scale. Quantitatively they are understood using the Heisenberg uncertainty principle. Let's say that in the lab you created the best possible vacuum with the strongest possible pumps and removed every atom from some volume inside a container. It's not physically possible, but we can imagine doing it. Even then, inside your container at the subatomic level so-called virtual particles are continuously coming into existence and then vanishing on a timescale inversely proportional to their energy. The vacuum is roiling with activity. While this may seem a bit out there, the effects of the roiling vacuum on atoms, after having let some back into your container, can be computed and measured to high accuracy. Quantum fluctuations are a well-established phenomena in laboratory experiments.

THE STANDARD MODEL OF COSMOLOGY

The model is that quantum fluctuations in the primordial energy density were stretched out to cosmic scales through the inflation of space. The fluctuations in the primordial field are now seen as the gravitational landscape that produced the hot and cold spots in the CMB. This means that when we look at the CMB we are looking directly at a manifestation of quantum processes. The random distribution in space of the hot and cold patches is a result of our quantum origins. We usually associate quantum processes with taking place on an atomic or subatomic scale. This is still true; it is just that inflation expands space so much that the quantum scale becomes the cosmic scale, a mind-blowing concept.

The expansion in inflation is similar in character to that for the cosmological constant discussed earlier, but in inflation the pressures are much much greater. Perhaps the origin of the processes is related. We don't know. Also, it may be that inflation is not the correct paradigm. It is possible that the universe goes through cycles of expansion and that we are in just one of the cycles. Even in this case, though, the origin of the CMB anisotropy can be traced to quantum fluctuations.

Let's revisit the maps of the CMB anisotropy in plate 7. We can now look at them with a new perspective. These maps show quantum processes now writ large across the sky. It is as though the evolution of the universe acts like a microscope to show us our quantum origins.

4.3 Pulling It All Together

While explaining the CMB has guided our path to an understanding of the universe, there are many other ways to study the universe. Cosmology is a broad field. The physics brought to bear includes everything from general relativity, to thermodynamics, to elementary particle theory. Observations are made in nearly every wavelength regime accessible to measurement and with state-of-the-art particle detectors. The observations come from nearby and from the farthest reaches of space. All of this evidence and theory is encompassed in the surprisingly simple standard model. Before summarizing the model, we touch on two major frontiers we have not discussed in much detail.

The most time-honored approach to cosmology is through observations of galaxies. As we saw, this was how Hubble and Lemaître pointed out that the universe was expanding. In addition to telescopes like the Hubble Space Telescope that can peer deeply and with high resolution in a given direction, there are others that measure the properties of millions of galaxies over more than a third of the sky. Probably the best known is the Sloan Digital Sky Survey. From this and related efforts we now have maps of the three-dimensional distribution of galaxies throughout much of the observable universe. We see in detail how galaxies clump. We can see how light from distant galaxies is bent on its way to us by the curved space near the intervening galaxies. By averaging over large volumes we can even see that there is a characteristic size for the clumping of galaxies that corresponds to the average CMB hot spot and cold spot sizes in plate 8a. For galaxies, the special spot size is called the "baryon-acoustic oscillation scale." To emphasize, the signature of the physical processes that produced the hot and cold spots in the CMB is also detected in the distribution of galaxies.

Quite independently of the galaxy and CMB observations, cosmologists have worked out the nuclear physics of the first three minutes of the universe in a study called Big

Bang Nucleosynthesis. The inputs to the calculation are the CMB temperature and nuclear interaction rates as measured in laboratories. The outputs are the abundances of the lightest elements: hydrogen, deuterium, helium, lithium, and beryllium. The first atomic nucleus to form was the deuteron, the nucleus of deuterium. It consists simply of one proton and one neutron. Before about 100 seconds, whenever a deuteron tried to form, it would be split apart by energetic photons corresponding to more than a billion kelvin. By 100 seconds, the universe had expanded and cooled enough for the forces that bind the proton and neutron to overcome the collisions with the photons that were trying to tear them apart. Thus, the deuteron could survive intact. In roughly the next 100 seconds, the deuterium was converted to helium through a series of nuclear interactions. By 1000 seconds, the other light nuclei formed. The process was a competition between the force that binds the neutrons and protons, the photons losing their energy because of the expansion of the universe, and the 10-minute decay time of the neutron.

The main predictions of the nucleosynthesis calculations are the overall cosmic fractions of atoms. As you might imagine from the above, there is an intimate relation between the energetics of the photons and the number of nuclei that are made. For the predicted nuclear abundance to match observations, there must have been about two billion photons per proton. These photons are of course the CMB.

The calculations predict that the atoms in the universe are primarily hydrogen (75% by mass) and helium (25% by mass) with only trace amounts of the other elements. Not only is this observed on the cosmic scale, but the Sun is 75% hydrogen and 25% helium. The calculations show that elements heavier than beryllium could not have been formed in the early universe. In general, the measurements of cosmic abundances of the light elements are in agreement with what one would infer from the CMB with one exception—lithium. Less is found than predicted. It is likely that the early stars eat the lithium but the mismatch between expectations and measurement may be telling us that there is an element of the calculation or model we are missing.

We now summarize the six parameters[3] of the cosmological model. The particular values we give result from fitting the gray curve in figure 3.3 to the CMB data. The values don't shift much, and the uncertainties improve, when additional datasets, such as the distribution of galaxies, are combined with the CMB. We give the specific symbols for the parameters as they are often encountered in the scientific literature.

As a foundation, the model stipulates that the universe is geometrically flat. We showed earlier that we could use measurements of the CMB and Hubble's constant to demonstrate that the geometry was flat. Indeed, when we do the calculation the result is consistent with flatness. However, instead of making the geometry part of the fit, we assume geometrical flatness and deduce Hubble's constant. This gives us the opportunity to compare the Hubble constant derived from the CMB, that is from the early universe, with the Hubble constant directly measured from the recession of galaxies versus distance. While the agreement is very good, it is not perfect. This too may be telling us that there is an element of the model we are missing—an exciting prospect—or that, perhaps, there are systematic errors in the measurements. The jury is out. Fortunately there are

3 Some advocate that the temperature of the CMB, 2.725 K, should be included in the list of parameters for a total of seven.

other ways to check the cosmic geometry, and they also tell us it is flat to the limits of measurement.

The first three parameters tell us about the contents of the universe. They are specified as fractions of the total, like the components in a typical pie chart as we discussed earlier.

1. Atoms account for about 5% of the universe. In the CMB anisotropy spectrum, figure 3.3, the ratio of the height of the first to the second peak gives a measure of the density of atomic nuclei in the early universe. It is not obvious that this should be the case and was only understood after we knew how to compute the curve in the figure. The value from the CMB anisotropy agrees with the value from Big Bang Nucleosynthesis. The fact that the stuff of which we are made accounts for just 5% of the net cosmic energy density gives a new perspective on our place in the universe. We specify this fraction with the Greek letter omega and say $\Omega_{atoms} = 0.05$.

2. Dark matter accounts for 25% of the universe. In the CMB anisotropy spectrum, the ratio of the height of the first to the third peak gives a measure of the dark matter density. This too is not obvious and was only understood after we knew how to compute the gray curve in figure 3.3. What is remarkable is that the amount of dark matter derived from the CMB anisotropy agrees with the value deduced from observations of the motions of stars and galaxies discussed in section 2.2, but the value from the CMB is much more precise. In addition, because the CMB comes to us from the decoupling era, the third peak tells us that dark matter existed in the early universe.[4]

---

4   Although the matter fraction of the total cosmic density changes with time, the ratio of atomic matter to dark matter was fixed well before decoupling.

CHAPTER FOUR

There must be new fundamental particles in Nature from the Big Bang that have never been detected in the lab. To specify the fraction of the universe that is dark matter, we write $\Omega_{DM} = 0.25$. What's more, we see that the stuff of which we are made accounts for just one sixth the total mass in the universe.

3. The cosmological constant accounts for 70% of the universe. We don't know what it is, but we have directly measured its presence through the cosmic acceleration. In the CMB, we determine it from the position of the first peak in figure 3.3. The value from the supernovae observations agrees with the value from the CMB. We write $\Omega_\Lambda = 0.70$.

There are of course other components such as the CMB radiation itself and the mass fraction of neutrinos. We know they are there but they are not significant enough that, with the current level of precision, they need to be included in the overall budget.

The next parameter is the most astrophysical one. It captures our rather scant knowledge of the entire complex process of the formation and subsequent explosion of the first stars, and the formation of the first galaxies. The intense light from these early stars and galaxies broke apart the hydrogen into its constituent protons and electrons, reionizing the universe. With the current level of precision, we need just one parameter to account for what no doubt we will one day find to be a rich process.

4. In the process of reionization, about 5–8% of the CMB photons were rescattered. In the analogy used to describe decoupling, it is as though a bit of fog rolled

in. Not too much—you could still see a distant shore—
but the visibility would not be perfect. The symbol that
describes the scattering is τ and is called the "optical
depth." We measure $\tau = 0.05 - 0.08$. But τ cannot be
determined with the temperature anisotropy alone. It
takes a measurement of the *polarization* of the CMB, a
topic we have not discussed. Polarization, along with
intensity and wavelength, is one of the three characteristics of a light wave. The polarization specifies the
direction in which the light wave is oscillating. For
example, light reflected off the hood of your car is horizontally polarized. That is, the light wave oscillates back
and forth horizontally. Polarized sunglasses block this
oscillation direction and its associated reflected glare.
Similarly, the electrons freed up in reionization scatter and polarize the CMB. If you could look at the
CMB with polarized "sunglasses" it would look slightly
different. In figure 3.3 reionization causes an overall
suppression of the spectrum, with just slightly more suppression at the largest angular scales. The optical depth
is the least well known of the cosmological parameters.

The next two parameters characterize the seeds of the fluctuations that gave rise to all the structure in the universe. The
concepts underlying these are beyond the scope of this book,
but we include them for completeness. These seeds led to the
CMB anisotropy spectrum and to the fluctuations in the total
matter in spheres of 25 million light-years in diameter that
were discussed in section 1.1. These primordial fluctuations
are described by the "primordial power spectrum." This is
similar in character to the CMB anisotropy power spectrum
(figure 3.3), but instead of describing the decoupling surface,

it describes fluctuations in density in three-dimensional space. As we look around the cosmos today, the three-dimensional fluctuations in density are large. In some places there are galaxies, in others clusters, and in others almost nothing. Before there were identifiable objects, the density fluctuations were much smaller. As we discussed, at decoupling the contrast was a part in 100,000. We use the primordial power spectrum to quantify the density fluctuations back at the beginning of the cosmic expansion.

5. The amplitude of the primordial power spectrum is encoded in the formidable symbol $\Delta_{\mathcal{R}}^2$. If we had a complete model of the universe that began with the quantum fluctuations and predicted, say, the fluctuations in matter in spheres of 25 million light-years diameter, we could relate $\Delta_{\mathcal{R}}^2$ to the rest of physics and its value would be known. Unfortunately, while we have a very successful framework, we do not yet know all the connections and so require it as a parameter.

6. The final parameter, called the "scalar spectral index" or n_s, is the most difficult to understand, but is also our best window into the birth of the universe. Like $\Delta_{\mathcal{R}}^2$, it tells us about the primordial fluctuations. In contrast to specifying the overall amplitude, it tells us how the primordial fluctuations depend on angular scale. To better grasp this, let's go back to the musical analogy that we used to describe figure 3.3. For a moment, let's put aside those peaks and troughs in the spectrum, from which we learn so much, and imagine the plot representing "white noise." In this case, the data points would lie along a flat horizontal line. All frequencies (all angular scales) would have the same loudness (or variance

as measured on the *y-axis*). The parameter n_s allows us to distinguish between "white noise" and, say, "pink noise," in which the bass notes have a somewhat greater loudness than the treble notes.[5] Using the CMB, we find that the primordial fluctuations, the "seeds," were ever so slightly larger in amplitude at large angular scales than they were at smaller ones. That is, the primordial cosmic noise is slightly pink.

When the process of cosmic structure formation was originally being studied, the scalar spectral index was argued to be unity, or $n_s = 1$, on general grounds. It was called the Harrison-Peebles-Zel'dovich spectrum after its authors. This value corresponds to white noise. Then, in the early 1980s, it was realized by Viatcheslav Mukhanov and Gennady Chibisov that this quantity could be computed from quantum principles operating as the universe was being born. We now know this index differs from unity by about 5%, that is $n_s = 0.95$, corresponding to just slightly "pink." This is the evidence that all the structure in the universe arose from quantum processes operating at a time when the universe was so compact and energetic that no known particles yet existed.

With these six parameters we can compute the properties and spectrum (the gray line in figure 3.3) not only for the CMB but for *any* cosmological measurement. We can

5 While our analogy is a reasonable representation, in practice n_s applies to the full three-dimensional primordial power spectrum and not the two-dimensional version of it that characterizes the CMB anisotropy. Also, there is a subtle convention, beyond our scope, that specifies when the loudness of a given pitch is defined. Last, for the experts, this use of the term "white noise" refers to the CMB power spectrum as plotted, as opposed to a constant C_ℓ.

CHAPTER FOUR

compute the age of the universe. The single most constraining observation is the CMB anisotropy, but the model is consistent with all measurements. In short, no matter how we look at the cosmos—with galaxy surveys, through exploding stars, through the abundance of the light elements, through the speeds of galaxies, or through the CMB—we need only the six parameters given above and the physical processes we explained in the preceding sections to describe what we observe.

In 1970, Allan Sandage wrote an article for *Physics Today* entitled "Cosmology: A search for two numbers." We now know it takes six, but with them we can account for more than Sandage thought possible. What does it mean to be able to describe something so simply and quantitatively? It means we understand how the pieces fit together, all the ones discussed in chapters 1–3 and more, to form a whole. We understand some deep connections in Nature. It means we can be proved wrong, not by different arguments but by a better quantitative model that describes more aspects of Nature. There are few systems studied by scientists that can be described so simply, completely, and with such high accuracy. We are fortunate that the observable universe is one of them.

THE STANDARD MODEL OF COSMOLOGY

CHAPTER FIVE

FRONTIERS OF
COSMOLOGY

THE STANDARD MODEL OF COSMOLOGY IS SO SUCCESSFUL that it is now a foundation from which we can look for departures. Through more precise measurements of the CMB we will learn, for example, the total mass of neutrinos. We might find that there are remnant gravitational waves from the birth of the universe that now pervade the universe. We could perhaps find that the cosmological constant isn't constant or that general relativity needs modification. Perhaps the universe isn't quite geometrically flat. Perhaps the fluctuations have a slightly different form and spectrum than we now measure. Perhaps a new particle in the early universe will reveal itself. To be able to determine any of these, we need more precise data. Before we touch on five especially active or promising frontier areas—the neutrino mass, gravitational waves, fundamental physics from structure formation, finding clusters of galaxies, and searching for subtle variations in the

CMB's temperature spectrum—we will first introduce a new observational technique: CMB lensing.

The image of the Bullet Cluster (plate 6) shows a clear separation of dark matter from normal matter. The location of the dark matter was found through an analysis of the gravitational lensing of far-distant galaxies by the Bullet Cluster itself. The cluster acts as a lens. We can take this to another level. The Bullet Cluster is a lens not just for the distant galaxies but for everything behind it, including the CMB. If we could measure the CMB with high precision right around the Bullet Cluster, we would see that it is distorted. One advantage of using the CMB as a backlight is that it comes from one surface at a precise distance, so the effect of lensing may be computed accurately. The Bullet Cluster is pretty massive so it stands out, but *all* the mass concentrations between us and the decoupling surface act as lenses. No matter where we look, the CMB is lensed. The effect is small, but with the current high-sensitivity instruments, it is readily identifiable.

How can we distinguish the underlying CMB anisotropy from the lensed version if it is lensed no matter where we look? The lensing has a distinctive effect on the CMB. It distorts the anisotropy in a special and calculable way. If you viewed the world through slightly textured glass, and you knew the characteristics of the texturing, you could figure out its effects on what you were seeing. In cosmology, the analog of the textured glass is the distribution of matter between us and the decoupling surface, and "the world" is the CMB.

There is a beautiful and deep connection here. Our model posits that the primordial power spectrum gave rise to the CMB anisotropy and to the fluctuations in matter throughout the volume of the observable universe interior to the decoupling surface. If we have the correct picture, we should be able to compute the CMB lensing to high accuracy because we know all the pieces of the puzzle. So far, the lensing of the CMB matches the prediction. This gives us added faith in the standard model because the predictions were made well before the measurements. The lensing measurements have an additional benefit. Similar to the way in which lensing by the Bullet Cluster tells us where the mass is located, lensing of the CMB gives us a two-dimensional projection on the sky of the distribution of dark matter throughout the universe. Maps of the mass distribution are already being made. We expect to continue learning more through CMB lensing in the future. The technique will play a large role in pursuing the first four frontier areas we address now.

Neutrinos. We have already mentioned neutrinos a few times in this text. Until recently they were thought to be mass-less. We now know they have to be more massive than one ten-millionth the electron's mass, but less than ten times that. Because there are so many in the universe, about 300 per cubic centimeter, they affect how cosmic structure grows. There are a number of ways they affect the CMB, but one of the most distinctive is through lensing.

If neutrinos are on the light side of the possible mass range, they act somewhat like photons, traversing the universe without affecting the matter distribution. If they are on the heavy side, they still travel quite fast and reduce the degree of clumping in the matter distribution by in effect transfer-ring mass from high-density regions to lower-density regions. The more massive the neutrino, the more the contrast is red-uced. The degree of clumping affects the lensing of the CMB because it's the fluctuations in matter that are producing the lensing; thus the more massive the neutrino, the smaller the lensing signal.

CHAPTER FIVE

The CMB lensing measurements are not quite sensitive enough to see this effect, but they will be soon. They are also not as informative as a laboratory measurement with regard to the defining characteristics of neutrinos. Primarily what we learn from the CMB is the neutrino's gravitation effect on the distribution of matter. The CMB observations can't distinguish between, say, the different neutrino types or other fundamental properties. Still, it would be amazing to determine one of the fundamental properties (the mass) of these most elusive of particles through their gravitational lensing of the CMB. We know so little about them that we might be surprised by what we find.

Earlier we mentioned that neutrinos, as we understand them, couldn't be the dark matter. We can now see why. If they act the way we think they should, they will stream out of the more dense regions and reduce the formation of cosmic structure. We'd see this in the distribution of galaxies, and we do not. Upcoming surveys of galaxies will be so sensitive that they will be able to see the neutrino's effects on cosmic structure. There is an opportunity to compare the neutrino's effect on the CMB and on the distribution of visible light. This is one of the many ways in which the cosmos is becoming a laboratory. There are a myriad of interlocking observations so that the deductions from one individual measurement can be compared against those of another.

In addition to the mass, with the CMB we have already begun to constrain, independently of laboratory measurements, the number of neutrino families. Improved measurements will lead to improved constraints. We might even find that there is a new kind of neutrino, or related particle, that we have not yet seen in nuclear reactions.

FRONTIERS OF COSMOLOGY

Gravitational waves. In many variants of the standard model, a background of gravitational waves is produced in the early universe. They are another form of the quantum fluctuations. In general these waves are a distortion of space and time that propagate across the universe at the speed of light. If a gravitational wave were aimed at a 100 cm by 100 cm plate, then in one half cycle it would shrink the width and expand the height. A half cycle later it would shrink the height and expand the width. If the change in height was 1 cm, we would say the strain is one part in a hundred, or 1%. The Laser Interferometer Gravitational-Wave Observatory (LIGO) detector on Earth detected gravitational waves from a pair of in-spiraling and merging black holes that were about 1.2 billion light-years away. The strain they measured was a part in 1 followed by 21 zeros. This is equivalent to detecting a change in distance between us and Proxima Centauri, the nearest star, which is 4.3 light-years away, with the precision of the width of a human hair. That is a staggeringly precise measurement.

The Big Bang might produce similar waves, in the form of "standing waves," but with wavelengths ranging in size from about 1% up to 100% the size of the observable universe. Since the wavelengths are so large, the distortions produced by the waves appear stationary to us. Some current models predict the strain should be about a part in 100,000. This is a far greater strain than detected by LIGO. It corresponds to measuring the height of a human to the width of a human hair.

Gravitational waves affect the anisotropy as well as the polarization of the CMB. By stretching and squeezing space, the gravitational waves subtly alter the CMB. The effect is so small that it can't be distinguished from the anisotropy produced by the primordial power spectrum. However, the

CHAPTER FIVE

gravitational waves affect the CMB polarization in a characteristic way. If we think of the polarization direction as represented by a series of short sticks, primordial gravitational waves impress on them a faint swirly pattern called a "primordial B-mode." Imagine you threw the contents of a box of round toothpicks on a large black floor so you could see them from the top of a short ladder. You'd want to throw them forcefully enough so that none of the toothpicks overlapped. Let's say that the orientation of the toothpicks represents the direction of the CMB polarization against the background sky. You then take a picture of it from atop the ladder. The pattern looks random. Now you look at that same pattern of toothpicks in a huge mirror and take a second picture. The last step is to line up the two pictures, the one taken directly of the floor and the one of the mirror image, and subtract them. The part of the first picture that goes away through the subtraction is called an "E-mode" and the part that remains is a "B-mode." In the standard model, the CMB polarization is almost purely E-mode: it looks the same in the mirror. So far, there are no traces of primordial B-modes.[1]

A detection of a primordial B-mode would be very exciting. It would provide a new and deep connection between the quantum regime of the very early universe and gravity. It would also provide a new test of fundamental theories of physics when they are extrapolated energies far beyond what can be achieved in an Earth-bound laboratory. If inflation is the correct model of the very early universe, a detection of gravitational waves might be just around the corner. In fact, for the original versions of inflation we should have seen them

1 Primordial gravitational waves produce E-modes and B-modes equally, but the E-modes are not as readily distinguished from the rest of the CMB as the B-modes.

already. A detection would also have strong implications for cyclic cosmological models. As they are currently understood, cyclic models cannot produce primordial B-modes at a level we might ever hope to measure with the CMB. A detection would rule them out.

To give a sense for how advanced the measurements have become, B-modes have been detected in the CMB. However, they are not from primordial gravitational waves. Rather, they are from the gravitational lensing of the E-modes! The same lensing effect that distorts the anisotropy also alters the CMB polarization. Just as with the lensing effects in the anisotropy, the lensing of the E-modes is at the predicted level, giving us even more confidence that we have the correct model of the universe.

Structure formation and basic physics. It is one thing to specify the contents of the universe. It is quite another to be able to understand how those ingredients combine and work together over billions of years to produce the universe we observe today. By carefully measuring how mass assembles over the ages we can test to see if the cosmological constant is indeed constant with time.

One way to approach the challenge is through a combination of galaxy surveys and the CMB. There are a number of surveys both in space and on the ground that will start to produce massive compendia of galaxies and their characteristics in the next decade or so. The largest on the ground will be the Large Synoptic Survey Telescope. It is expected to measure more than 10 billion galaxies over almost half the sky. Over the same region, deep surveys will be made of the CMB made from the ground. The gravitational lensing signal from both the galaxy surveys and the CMB will be particularly exciting to compare. There are many other ways to combine the data as well. What we expect to emerge is an exquisitely detailed three-dimensional picture of the universe. With detailed interlocking datasets, we can look for tiny departures in the expansion rate versus time from the predictions for an unchanging cosmological constant.

The Sunyaev-Zel'dovich (SZ) effect and clusters of galaxies. The largest gravitationally bound objects in the universe are clusters of galaxies. They are individual identifiable systems made of hundreds to many thousands of galaxies with names like the Virgo Cluster, Coma Cluster, or, as we saw earlier, the Bullet Cluster. A typical cluster is six million light-years across, about 60 times the size of the Milky Way. If towns and villages on a map are like galaxies, clusters are like major cities. One of the characteristics of clusters is they are full of hot gas that is not bound up in stars. This gas emits X-rays, as we saw in the example of the Bullet Cluster.

Rashid Sunyaev and Yacov Zel'dovich pointed out in the 1970s that the hot gas in clusters affects the CMB. The gas is so hot that it is ionized and in essence composed of free protons and electrons. When a CMB photon on its way to us from the decoupling surface interacts with an electron in hot cluster gas, it is scattered. The hot electron gives the photon some of its energy. This alters the spectrum of the CMB shown in figure 2.1, in effect taking energy out of the part with wavelengths longer than one and a half millimeters and putting it at shorter wavelengths. In other words, the scattering distorts the spectrum of the CMB.

This means that if we scan the skies at wavelengths longer than one and a half millimeters, clusters will appear colder than 2.725 K. The dip in temperature is around a thousandth of a kelvin, so with modern detectors it is quite easy to see. More than a thousand clusters have been seen using this characteristic "SZ" signature in the CMB, and before long there will be ten times this many.

One of the features of the SZ signature is that it is almost independent of when the scattering took place. For the same temperature electrons, when the universe was more compact, the CMB was hotter and the corresponding dip in temperature greater. The expansion of the universe cools the scattered photons just as it does with the CMB so the net SZ effect stays the same size. With the SZ effect we can look out to great distances, back to a time when clusters were forming. In a given direction, we can measure all the clusters in the observable universe above some mass limit. We can then see how many there are versus time and compare that to predictions of structure formation. The clusters give us another way to probe the cosmological constant.

Clusters highlight another example of the important link between different types of observations. With the SZ effect, we can't tell the distance to a cluster or its mass. We need to observe in the visible or infrared to see how far away one is. There are a number of ways to determine a cluster's mass. Probably the best way is to use gravitational lensing as was done with visible light for the Bullet Cluster. Soon we will have large catalogues of clusters complete with distances and masses and thus another way to look for new elements in the standard model.

The temperature spectrum. We noted earlier that if the source of radiation is a blackbody, all you need to do is specify its temperature and you know the intensity at all wavelengths. To the limits of measurement we know the CMB is a blackbody (away from clusters of galaxies!). Put another way, it is described by the Planck function as shown in figure 2.1. However, if the source is not a blackbody, then the effective temperature depends on the wavelength. This could be the case if there was a large injection of energy during the cosmic evolution, say from the decay of some particle, or if the universe evolved in such a manner that the radiation did not have time to come into equilibrium with the particles. There are a number of known processes that should alter the temperature spectrum at levels a little greater than a factor of ten below current limits. They include the reionization associated with the formation of the first stars and the spectral distortion produced by the SZ effect of the combination of all groups of galaxies and clusters. These signals are too small to detect with the current experimental methods, but instruments are being designed to search for them and other features.

Summary and Conclusions

Let's take stock of the major themes we've discussed. In chapter 1 we got a sense of the almost overwhelming vastness of the universe. The Milky Way is just a speck of dust in the cosmic expanse. And recall, it contains 100 billion stars and most of those have planets. From a cosmic point of view, Earth is insignificant. A starting point for making quantitative sense of the vastness is Einstein's Cosmological Principle. When the principle is meshed with observations, we find that if the universe is averaged over 25 million-light-year-diameter spheres, it more or less looks the same no matter where we are. In other words, from a coarse-grained perspective, the universe is homogeneous.

We also saw that this whole vast landscape is expanding. What's more, the expansion is accelerating. Again, these are observations, not theories. This is the way the universe acts.

One way to think about the observations is that space itself is expanding and that the contents are going along for the ride. But we don't know why space would expand. It is just a description. By extrapolating the expansion back in time, we realized that there was a beginning a finite time ago, in fact 13.8 billion years ago. It may not have been the beginning for all existence and for all time, but it was the start for our observable universe. Knowing that the speed of light is fixed made us realize that the farther out in space we peer, the farther back in time we look: remember, telescopes are like time machines. When we look back far enough, we see the CMB.

In chapter 2 we took a tally of the major constituents: the CMB, the atoms, the dark matter, and the cosmological constant. We know there have to be more components—for example there should be neutrinos—but they are subdominant enough that the model doesn't need them to account for the observations. The atoms are conspicuously clumped into galaxies; these are our cosmic signposts. The dark matter distribution is more puffy than that of the atoms, but it is still clumped. The energy density associated with the cosmological constant suffuses space. As far as we have been able to measure, it is not clumped. The CMB also suffuses space, but its energy density is insignificant compared to that of the atoms, dark matter, and cosmological constant.

After describing, in chapter 3, how we measure the CMB and reduce the anisotropy maps to a usable form, we turned to the interpretation of the data in chapter 4. Going back to the beginning of the book, perhaps the most amazing thing about the universe is that we can understand it at its grandest scales to percent-level precision. In a nutshell, a hot, dense universe expanded from a fiery beginning that we call the Big Bang. Quantum fluctuations intrinsic to the fabric of

CHAPTER FIVE

the primordial spacetime, and expanded by the rapid early expansion, grew into fluctuations in the strength of gravity throughout space. The CMB gives us a two-dimensional snapshot of these fluctuations some 400,000 years after the Big Bang. As the universe evolved, the dark matter and atoms responded to the variations in gravity to eventually form all the structure in the universe. The cosmological constant, initially inconsequential, now drives an accelerated expansion and will increasingly dominate the universe.

It is remarkable that humankind has arrived at the standard model of cosmology. We cosmologists feel fortunate to have been alive in the decades when the explosion of knowledge about the universe took place. Most of us in the field recall a time when we did not know the geometry of the universe, its contents, or its age. As the data have become more and more precise, whole classes of cosmological models have been shown to be wrong. As we have emphasized, it is precise measurements that are the foundation of the standard model. The dramatic advance in cosmology has occurred through the ability to compare models to measurements. It turns out that the early universe is simple and that the physics that describes it is straightforward. It did not have to be this way, but Nature was kind in letting us learn so much.

We now have a powerful and predictive model, but even within that context there are still many open questions. Some we can address with better measurements or deeper theories: What is the dark matter? Why is it that our universe is predominantly matter as opposed to a combination of matter and anti-matter? What is the physics of the very earliest times? What is the cosmological constant telling us about the vacuum? As far as we know there is no particular "need" for it. Although we talk about "expanding space," we do not really know what space is. The clues may well be all around us but we haven't thought of them in the right way. There are questions we may never be able to answer conclusively: Are there multiple universes? Are we in just one of an endless series of cycles?

The cosmos has captured the imagination of humans since time immemorial. Although the recent advances are dramatic, the quest for ever deeper knowledge on both theoretical and experimental fronts continues. For those observing the cosmos, nothing is more exciting than finding something new, or learning that one of the elements of the standard model needs to be considered in a new light. There is a huge amount left to be learned from the CMB, and we will likely be measuring it for years to come.

CHAPTER FIVE

APPENDIXES

A.1   The Electromagnetic Spectrum

FIGURE A.1 SHOWS THE ELECTROMAGNETIC SPECTRUM OVER a wide range of wavelengths. The units on the x-axis change from centimeters (cm) on the left to microns on the right to connect with the text. The scales merge at $0.1\,\text{cm} = 1\,\text{mm} = 1000\,\text{microns}$. Note that the wavelength gets smaller going to the right, which means that the energy of a photon increases going from left to the right.

Channel 83 on your TV, which is not in general use, has a wavelength of 34 cm. Microwave ovens operate at 12.2 cm. These are indicated as lines because most of the energy is concentrated near one wavelength. The cosmic microwave background (CMB) is a blackbody emitter that peaks near 0.1 cm but emits power over a large range of wavelengths. This is the same spectrum as shown in figure 2.1 but now in a broader context. The next vertical line shows the wavelength for the

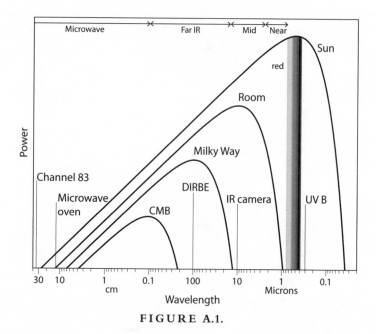

FIGURE A.1.

DIRBE image in plate 3. The spectrum labeled "Milky Way" corresponds to a blackbody at 30 K. The next spectrum is for a room temperature blackbody (300 K). IR cameras measure this thermal emission. The spectrum for the 6000 K Sun peaks around a wavelength of 0.5 microns. The gray scale corresponds to the colors of visible light our eyes detect and runs from red on the left to violet on the right. The UV spectrum is found at slightly shorter wavelengths. UV B radiation is at 0.3 microns. You can see that the Sun is still quite intense there, but we can't see the UV light. Along the top of the plot are the designations for the "Microwave" wavelength band, "Far-infrared" band, and "Mid" and "Near" infrared bands. You can also see that the peaks of the four blackbody spectra follow the Wien displacement law.

A.2 Expanding Space

"Expanding space" is a controversial phrase. We use it simply as an intuitive description of the change in the scale of the universe with time. We take guidance from Einstein: "In that sense one can say, according to Friedmann,[1] that the theory demands an expansion of space." And "It is indeed an exacting requirement to have to ascribe physical reality to space in general, and especially to empty space."

The coordinate system we use to measure the locations of objects in the universe is unambiguously expanding. At the same time, for most of the age of the universe, there is nothing that pushes galaxies apart that the phrase "expanding space" might bring to mind. Gravity is only attractive. The evolution of the universe during this time, over regions larger than shown in figures 1.3 and 1.4, can be described by giving the galaxies initial velocities and computing how they interact under the force of gravity.

However, for the past 4 billion years, since the cosmological constant has become the dominant form of energy density, a new force has come to dominate the universe that does indeed push galaxies apart. That force is quantified with the cosmological constant. Its action can be described as "expanding space" or "making space." Similarly, if inflation is the correct model for the early universe, it too can be described as "expanding space," but at an exponential rate over a very brief time. During inflation there is a force that pushes particles apart that is much stronger than gravity. The source of

1   Alexander Friedmann derived the equations that describe cosmology from the general theory of relativity as discussed in section 2.3. The quotes are from *Relativity* by Albert Einstein, Crown Publishers, 1961.

this force is an effective cosmological constant that is much larger than the one we currently observe.

There is another example of "making space." If the universe were described by a closed geometry, corresponding to the righthand image in figure 4.1, the volume of the universe would be finite and would change with time. Space would indeed be created.

Understanding the nature of space—the nature of the vacuum—is at the forefront of physics. We do not understand the vacuum at a deep level. For some situations, we are almost forced to think of space as expanding, whereas for others the expansion of space may lull us into thinking there are forces that don't exist. Nevertheless, we find the concept of an expanding space useful for envisioning many aspects of the universe.

A.3 Table of the Cosmic Time Line

Age	Compactness or Scale Factor	Event
0	Minuscule!	Our definition of the "Big Bang." §1.2, §1.3
1.4×10^{-14} sec	2.2×10^{-17}	Typical energy of a photon equals the particle interaction energy at the Large Hadron Collider. §2.1
0.000025 sec	1×10^{-12}	Quark-gluon plasma as seen at RHIC. §2.1
3 min	3×10^{-9}	The nuclei of H, He, Li, and Be formed. The temperature was 1 billion K. §2.1, §2.4, §4.3

continued

Continued

Age	Compactness or Scale Factor	Event
1 year	1×10^{-6}	Appendix A.4.
51,000 yrs	0.00029	"Matter-radiation equality." The dominant form of energy density changes from radiation to matter and cosmic structure can start to grow. §2.4
400,000 yrs	0.001	"Decoupling." Hydrogen atoms form and the CMB is free to roam the universe. Some call this time "recombination." §2.4, §3.2
1 million yrs	0.0017	
200 million yrs	0.05	First objects form. §1.6, §2.4
370 million yrs	0.078	Most distant object yet identified. §A.4
0.4–0.7 billion yrs	0.08–0.12	Most distant objects in the Hubble Ultra Deep Field. §1.6, §2.4
0.5–1 billion yrs	0.1–0.15	"Reionization." The universe was reionized by the first stars and the free electrons scatter 5–8% of the CMB photons. §2.4, §4.3
5.9 billion yrs	0.5	Universe is twice as compact. §1.3, §1.6
9.3 billion yrs	0.71	Time when the Earth and Moon appeared. §1.3

continued

APPENDIXES

116

Continued

Age	Compactness or Scale Factor	Event
10 billion yrs	0.75	"Matter-Λ equality." The dominant form of effective energy density changes from matter to dark energy. §2.4
13.7 billion yrs	0.993	Dinosaurs roamed the Earth. §1.3
13.8 billion yrs	1	We live in a ΛCDM universe.

For the extremely small numbers, we have had to introduce scientific notation in which the exponent tells where to place the decimal point. For example, $1 \times 10^2 = 100$ and $1 \times 10^{-2} = 0.01$. The compactness is the number by which one should multiply the scale of the current universe to determine how much closer objects were in the past.

A.4 The Observable Universe versus Time

Figure A.4 shows the size of the observable universe versus its age. In going from left to right, the vertical dashed lines show when cosmic structure started to grow (section 2.1), when the CMB decoupled from the primordial plasma (section 2.4), and the "distance" to one of the farthest identifiable objects.

When we see a distant object, often the first question that jumps to mind is "how far away is it?" For the universe, we have to be especially careful in specifying *when* we want to know how far away it is. As light propagates to us from a distant object, the universe expands. By the time we receive the light, the universe has expanded. Although the "light travel distance" back to the Big Bang is 13.8 billion light-years, over that 13.8 billion years the universe has expanded a huge amount so its current "radius," called the "comoving distance" in scientific literature, or more popularly the radius

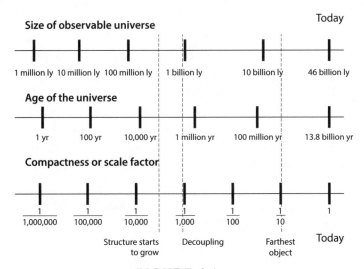

FIGURE A.4.

of the "observable universe," is 46 billion light-years. This corresponds to 92 billion light-years in diameter, about three times the size we gave in section 1.4.

The natural way to think about the universe is in terms of its compactness or "scale." We ask its age and size with regard to when it was 10 or 100 or a billion times more compact. The reason for this is that the fundamental physical properties—the temperature, the densities, the rate of expansion, etc.—depend on the compactness. Then, from the history of compactness, we deduce the age and size. For example, from the lefthand side of the figure we read that when the universe was one million times more compact, it was one year old, and the size of the observable universe was a couple of million light-years. At this time the CMB was one million times hotter because temperature is directly proportional to compactness.

One of the farthest identifiable objects is a galaxy called EGSY8p7. The light from it started on its way to us when the universe was about ten times more compact. The light we now observe was emitted when the universe was 0.6 billion years old and so has been traveling to us for $13.8 - 0.6 = 13.2$ billion years. In plate 5 this is in the purple band that the Hubble Ultra Deep Field can reach. The question of "how far away is it" isn't really the right question to ask because the universe has expanded so much since it emitted the light we just now observe.

We could have extended the lefthand side of the plot much further. Some earlier times and associated events are given in appendix A.3.

APPENDIXES

INDEX

age of universe, 18, 79
angular size, 6, 84
anisotropy, 58

B-modes and E-modes, 103
baryon acoustic oscillations, 89
beach balls as photons, 32, 34
big bang, 15, 20
big bang nucleosynthesis, 90, 92
blackbody radiation, 30, 112
Bullet Cluster, 43, 99, plate 6

clusters of galaxies, 57, 105
CMB hot spot size, 73, plate 8
CMB lensing, 99
CMB polarization, 30, 94, 102
CMB, where it comes from, 38
compactness, 17, 20, 114–116
COsmic Background Explorer
  (COBE), 4, 63
cosmic densities, 28, 50
cosmic triangle, 83
cosmological constant, Λ, 48, 98, 109
cosmological principle, 8, 22, 107
cosmological redshift, 36
cyclic model, 20, 88

dark matter, 29, 41
decoupling of CMB, 55, 69, 115
density, 17, 28

deuterium, 39, 90
Diffuse InfraRed Background Explorer
  (DIRBE), 4, plate 3
dinosaurs, 21, 28
Doppler effect, 35

expansion of space, 14, 113

Friedmann, Alexander, 50

galactic plane, 2, plate 2
gravitational lensing, 45
gravitational waves, 102

homogeneity, 8
Hubble Ultra Deep Field, 6, 22, 68,
  plate 4
Hubble's constant, 10
Hubble-Lamaître law, 10
humans, 21, 25
hydrogen ionization, 36
hydrogen to helium ratio, 52, 90

ice cream, 24
inflation model, 20, 86
isotropy, 8, 58

Kamiokande detector, 25

Lagrange point, 66
Large Hadron Collider, 37, 45

Large Magellanic Cloud, 25
Large Synoptic Survey Telescope, 104
Laser Interferometer Gravitational-Wave Observatory (LIGO), 102
light-years, 2
limitations to describing space, 15, 113
Local Group, 5

matter-Λ equality, 29
matter-radiation equality, 29
micron, unit of distance, 3
Milky Way, 2, 112
Mitchell, Joni, 56
Modified Newtonian Dynamics, 46
Mollweide projection, 58
Mukhanov and Chibisov, 96

neutrinos, 25, 40, 100
neutrinos as dark matter, 42

observable universe, 7, 23

Peebles and Yu, 73
Penzias and Wilson, 62
photon density, 34
Planck blackbody spectrum, 33
Planck satellite, 58, 64
primordial plasma, 52, 69, 72, 116
primordial power spectrum, 94

quantum fluctuations, 87, 108
quark-gluon plasma, 37

recombination, 115
redshift, 35
reionization, 57, 93, 107
Relativistic Heavy Ion Collider, 37
Rubin and Ford, 41

scalar spectral index, 95
scale factor, 21, 114–116
seeds of structure formation, 94
Sloan Digital Sky Survey, 89
standard model of particle physics, 45
sun, 2, 31, 112
Sunyaev and Zel'dovich, 73, 105
supernovae, 25, 47

Tinkertoys, 9, 21
topology, 82

UV radiation, 31, 112

vacuum, 16, 48, 87, 114

Wien displacement law, 36, 112
WMAP satellite, 58, 64

x-rays from clusters, 43, 105

Zwicky, Fritz, 41